The Politics of Trauma in Education

The Politics of Trauma in Education

Michalions Zembylas

palgrave
macmillan

THE POLITICS OF TRAUMA IN EDUCATION

First published in 2008 by PALGRAVE MACMILLAN® in the US - a division of St. Martin's Press LLC, 175 Fifth Avenue, New York, NY 10010.

Where this book is distributed in the UK, Europe and the rest of the world, this is by Palgrave Macmillan, a division of Macmillan Publishers Limited, registered in England, company number 785998, of Houndmills, Basingstoke, Hampshire RG21 6XS.

Palgrave Macmillan is the global academic imprint of the above companies and has companies and representatives throughout the world.

Palgrave® and Macmillan® are registered trademarks in the United States, the United Kingdom, Europe and other countries.

ISBN-10: 0-230-60576-1
ISBN-13: 978-0-230-60576-3

Library of Congress Cataloguing-in-Publication data

Zembylas, Michalinos.
The politics of trauma in education / Michalinos Zembylas.
p. cm.
Includes bibliographical references and index.
ISBN 0-230-60576-1
1. Education—Political aspects. I. Title.

LC71.Z42 2008
379—dc22 2008004021

A catalogue record of the book is available from the British Library.

Design by Westchester Book Group.

First Edition: June 2008

10 9 8 7 6 5 4 3 2 1

Printed in the United States of America.

For my daughter Mariza

CONTENTS

FOREWORD

I fear the state; I fear the nation-state more and the democratic state even more. I fear their liberal veil, their humanistic rhetoric and fear the most their enlightened compulsory free education law. Compulsory and free I read as an oxymoron, as I think it should be read. I realize "free" in the compound is to be read as a statement on the economics of education and not as a characterization of its nature and goals, and yet within the context in which the compound is built, I believe it to be, for the most part, confusing.

Etymologically, "free" points to a person not being bound or subject or enslaved; in the primary Germanic and Celtic sense it is a term of endearment (friend), having been applied as the distinctive epithet of those members of the household who were connected by ties of kindred with the head, as opposed to the slaves. Free also means to be exempt from payment, tribute, etcetera, but all these terms seem to collude, allowing little space for a consideration of freedom as "truly" free from payment (or commitment as some would call it). Add "compulsory" to this collusion and you are in more trouble than expected.

The development of mass (compulsory and free) education, through schooling, has been shown to be tightly related to the industrial revolution and the development of the nation-state (Bekerman & Silberman, 2003; Gellner, 1983; Smith, 1998). Both were in need of recruiting masses to their service; masses with basic cognitive and behavioral skills that could serve the needs of the nation-state and its economic structure. As such, schools are in no way disinterested arenas within which neutral knowledge or skills are transmitted from the minds of specialists to those of passive individuals. In the modern era, schools have served as the primary means by which sovereigns have unified the different local groups inhabiting the areas they were successful in subordinating to their power, under one flag, one language, and one narrative. Many freedoms have been necessarily denied to achieve this oneness—among others the freedom to think differently. In extreme times, the freedom to live had to be denied from the free citizen in the Free State, and the product of free education was disciplined to sacrifice his life for the benefit of the sovereign.

Nation-state is a powerful machinery loved by all; it runs free of ideological arguments and all—left, center, and right—are deeply attached to it. Nation-state is indeed beneficial to those in power irrespective of ideological attachments. All nations invent themselves and behind each one stands a "mediocre historian" who helps to engage the masses as born-again believers in the secular equivalent to the church that, through the ingenious manipulation of public ceremonies and the production of public monuments, molds a unifying history, a collective memory (Gillis, 1994; Hendelman, 2004). Israel, "my country," is no exception. Israel has also engaged filtering strategies in order to develop a narrative that serves its unification and justifies its primordial status as a "nation." While doing so it has, like many other states, violently silenced many other possible memories and many other possible voices; it has been very successful at this. One of the strategies adopted in achieving this goal has been an over emphasis on Holocaust studies/experiences—the most outstanding of which are the visits of Israeli school children to Poland. These visits have been carefully devised to inscribe on the Israeli youth a strong sense of belonging to the collective of the Jewish nation by procuring a merging of the present Jewish-Israeli collective with the victims of the past Holocaust. These trips are central to Israel's civil religion that fuses the individual's identity with that of the collective through integrative practices that offer legitimization to hegemonic goals, now seen as natural. They serve to mobilize the society's energies in pursuit of hegemonic goals, all geared towards the constitution (with the help of many other institutional practices) of children in the image of the "good, self-sacrificing, citizen."

Over 12 years ago, Shlomit, the second of my three daughters, decided she wanted to join the group that was going to participate in the visit to Poland in her school. She was at that time around 15 to 16 years of age, and it would be her first visit to a country abroad. I worried from the first moment. I thought I knew exactly why the school encouraged them to go but also knew that opposing it would not necessarily produce better results. I had to take into account a plethora of considerations. No human in his right mind can out-front deny the importance of learning about the Holocaust (Jewish and others); for Jews this is even more difficult, for Israeli Jews it is almost impossible. We have all been raised in spheres that leave no space for doubt. Holocaust is a rationale, a justification; Holocaust has become the foundation of Israel's existence and doubting its foundational character might be blasphemous. Moreover, for most Jews of European origin Holocaust involves some personal memory (mostly erased memories), and any hint at doubting its labor in immaculate memory brings us to doubt our own intentions, even when well trained in critical perspectives. We are indeed prisoners of a dialogue in which the hegemonic discourse almost always triumphs.

Shlomit knew something about my position; she had heard me express my doubts in the past yet I chose not to engage her in a new conversation before the trip. I chose instead to write a letter that I gave her and asked her to open some time later while in Poland. In the letter I shared with her my doubts and anxieties; my doubts regarding the place of the Holocaust in our lives, and my anxieties regarding the doubts I was expressing. I told Shlomit I was not sure we should think about the Holocaust as a Jewish problem. It was indeed a tragedy but not a problem, and if a problem at all, I told her, it seemed to me to be a German problem and not a Jewish one. I asked her to remember that the Poles were not the Nazis and that I found it curious that the Ministry of Education had chosen Poland as the site for pilgrimage, focusing on the dead (not only the Jews) and the killing technologies instead of on the murderers and the technologies (political, legal, economical, and military) they developed to commit their crimes. I told her that unfortunately there is no immunization against evil, and that given the circumstances we could all be Nazis. I asked her to be careful not to follow uncritically in the footsteps of leaders, for that was the beginning of evil. Last I told her I thought the most important thing Jews could learn was what they had known for centuries: that difference should be defended and respected (all difference but for the Nazi difference that is easily identified when present) and that we carry responsibility to prevent ourselves and others from adopting Nazi policies, so as indeed to have learned something from history. The letter seems to have worked, somewhat. When she returned she mentioned its reading as a relevant event in her visit and, if to judge from her development since then, she seems, at least, to confront similar conflicts to the ones I have described for myself above. I am happy my other two daughters seem to walk on a similar path. Yet, for most Jewish Israeli citizens, Holocaust is a learned/taught trauma, a learned witnessing that, at its best, only reflects the doxa of the establishment and the manifestations of hegemonic Zionist ideology.

Lyotard (1988) suggests that it is the nature of a victim not to be able to prove that he has been done wrong. Ophir (2000) suggests that from this perspective, a victim is unable, for lack of proof, to expect a replacement or compensation for his loss; thus the victim's loss becomes irretrievable. In a paradoxical way, being a victim can be immensely powerful. The victim's power derives from his ability to determine if there is anything that can be done to terminate his status as a victim. To retain this power, no proof of the wrong and thus no possible replacement or compensation should be found.

Israeli Zionist Jews, since their acceptance of the Holocaust as the ethos that substantiates the state's existence, have positioned themselves as the ultimate victim and thus denied (themselves and others) the option of finding a solution to the intractable conflict in which they find themselves with the Palestinian people. This statement should be in some ways qualified, for in this case "proof" of the wrongs done to the Jews in Europe is a demonstrable historical fact and Israel has acknowledged the possibility of being compensated by reaching a Reparations Agreement with West Germany. According to this agreement that was signed on September 10, 1952, and entered in force on March 27, 1953, West Germany was to pay Israel for the slave labor and persecution of Jews during the Holocaust, and to compensate for the Jewish property that was stolen by the Nazis.

Yet to this day Israel, mainly in dialogue with the Palestinian population but also in much of the dialogue it sustains with its partners and foes in international relations, emphasizes its position as victim. Sustaining its position as a victim has been a rather easy task, successfully achieved, as mentioned through the exploitation of the (nation) state's educational apparatus, an apparatus that by "nature" is already involved in the production of "grand narratives." The main characteristics of the cultural tools implemented in the educational system are lead by a self-reinforcing ideology that oscillates between the isomorphic oppositional rhetoric of "to remember" and "not to forget." The victim, free, now, from the need to recognize the possibility of being offered compensation, is also free from offering forgiveness and thus able to sustain his power that emanates from his victimhood. To stick to their position, present day Zionist Israeli Jews need a fixed remembering, which denies the "true" nature of remembering (always in oscillation with forgetting). Not surprisingly, forgetting and forgiveness are intimately connected in that both involve a directive of returning and of offering back. If the above is agreed upon, it follows that a fixed remembering, constitutive of a refusal to forget and allowing for a denial of forgiveness, helps sustain a (political/ideological) position that sustains conflict. Forgetting would allow for for-giving-back-the-conquered-lands that would/could bring about a partial solution to the present political stalemate. Yet there is an immediate need to qualify the necessity to forget by specifying the use of remembering.

When we point critically at remembering, we do not mean the remembering that honors/recognizes memory for the sake of weaving it into a narrative that substantiates our present existence. Accepting a material settlement with the German government does not necessarily mean that the Holocaust victims are forgotten; it just means that, within the complex and transient events of a developing life-story, spaces have been opened for a renewed dialogue (hopefully now fed by a memory that honors the past and guides us toward a better future). What we criticize are the rememberings geared toward the prevention of forgetting and forgiving.

It is at this point that Zembylas's book becomes necessary for it seeks ways to subvert uncritical responses to traumatic memories and narratives. Israelis are not the only ones who use traumatic strategies to sustain their rights: Greek Cypriots, North Irelanders of sorts, and others, many others, use or have used similar strategies. Zembylas's book forces upon us a critical approach to traumatic memories guided by the introduction of emotional ambivalence, as a central tool in the hands of educators, that can affect the politics of hatred and fear born out of traumatic remembrance in/and its political use. Such a process also offers the potential to deconstruct the positioned premised on the existence of clearly differentiated identities, thereby opening spaces for a substantial change in our engaging with alterity, no longer based on negation and duality but, rather open to dialogue. I know schools and their power too well to believe that change will come with ease. Still, if offered a chance, Zembylas's suggestions might help us to confront the homogenizing power of "massive education." The inquiry in this book is essential to educational efforts initiated to support peace processes. Developing relevant educational dialogues needs to overcome the stalemate implied in victimhood. Let us give Zembylas's suggestions a chance and see if we can benefit from them.

In one of his famous stories entitled *Funes the Memorial,* Jorge Luis Borges (1996) doubted whether Funes could think. Thinking has to do with forgetting, he thought, and forgetting was the one thing Funes could not do. In Funes's world, there were only immediate details that he could not forget. When flooded by memory, Funes lost his ability to think, to reflect; he drowned in almost immediate-past details and so he died, trapped in memory. If we seek to escape Funes's destiny, we need to do some forgetting, without which reconciliation and coexistence seem to be unattainable.

Zvi Bekerman
Hebrew University, Jerusalem

References

Bekerman, Z., & Silberman, M. (2003). The corruption of culture and education by the nation state. *Journal of Modern Jewish Studies, 2*(1), 19–34.

Borges, J. L. (Ed.). (1996). *Funes el memorioso* (Vol. 1). Barcelona, Spain: EMECE.

Gellner, E. (1983). *Nations and nationality.* Oxford: Basic Blackwell.

Gillis, J. R. (Ed.). (1994). *Commemorations: The politics of national identity.* Princeton, NJ: Princeton University Press.

Hendelman, D. (2004). *Nationalism and the Israeli state: Bureaucratic logic in public events.* New York: Berg.

Lyotard, J.-F. (1988). *The differend: Phrases in dispute.* Minneapolis, MI: University of Minnesota Press.

Ophir, A. (2000). The identity of the victims and the victims of identity: A critique of Zionist ideology for a post-Zionist age. In L. J. Silberstein (Ed.), *Mapping Jewish identities* (pp. 174–200). New York: New York University Press.

Smith, A. D. (1998). *Nationalism and modernism.* London: Rutledge.

Acknowledgments

This book grew out of the work I have been doing during the last few years on the history and the politics of trauma and reconciliation in Cyprus (my homeland) and the role of education in this context. I would like to thank the Open University of Cyprus for its support for this work. In addition, I would like to express my heartfelt thanks to the United Nations Development Program (Action for Cooperation and Trust) in Cyprus and CARDET, Cyprus (especially my very good friend Charalambos Vrasidas) for their support of my research and development projects in this area. I owe a great debt of gratitude to Zvi Bekerman and Claire McGlynn for their invaluable insights and their generosity; our discussions encouraged me to think more deeply about the multiple aspects of the politics of trauma in education. I am also indebted to Brigitte Shull and Julia Cohen from Palgrave Macmillan for their ongoing support throughout all the stages of the production of this book.

Finally, I want to thank the editors of the following academic journals for permission to quote from and rework previously published (or upcoming) material of mine: *Intercultural Education* ("The affective politics of hatred: Implications for education" Vol. 18, No. 3, 2007, pp. 177–192); *Journal of Peace Education* ("The politics of trauma: Empathy, reconciliation and education" Vol. 4, No. 2, 2007, pp. 207–224); *Teaching Education* ("Mobilizing anger for social justice in education: The politicization of the emotions in education" Vol. 18, No. 1, 2007, pp. 15–28); *Discourse: Studies in the Cultural Politics of Education* ("Trauma, justice and the politics of emotion: The violence of sentimentality in education" Vol. 29, No.1, 2008, pp. 1–17); *Curriculum Inquiry* ("Education and the dangerous memories of historical trauma: Narratives of pain, narratives of hope" [with Zvi Bekerman] Vol. 38, No. 2, 2008, pp. 125–154); *Education, Citizenship and Social Justice* ("The politics of shame in intercultural education" forthcoming).

Introduction: Trauma, Politics, and Education

Engaging with trauma in educational contexts is not an easy task. This book is about the ways in which trauma enters educational discourses and practices and about the need to explore its pedagogical, ethical, and political implications in education. This exploration begins with reference to a personal consternation, namely the division of my homeland, Cyprus, and the traumatic events impacting the lives of thousands of Greek-Cypriot and Turkish-Cypriot children for many years now. The story offers insight into some of the book's main themes, namely that traumatic events shape the lives of educators and students *affectively*, producing subjects of a "wound culture" (Selzer, 1997), that is, individuals who regulate their perceptions and contacts with others on the basis of the impact of trauma both on themselves and on their respective cultures. It is shown that the political-ideological context within which traumatic events take place shapes their impact and thus it is hard to separate individual and collective trauma (Kaplan, 2005).

However, it is shown that engaging with trauma in educational contexts is not only a difficult task but also a transformative form of pedagogical engagement, because the development of a critical history and politics of trauma in education—I term this work *critical emotional praxis*—creates openings for different affective relations with others. The aim of critical emotional praxis is to explore the conditions under which trauma impacts educators' and students' lives, to destabilize and denaturalize that regime of thought that perpetuates a conflicting ethos with those who are deemed responsible for "our" trauma, and to invent new practices of relating with others.

On this crucial point, my theoretical concerns invoke a mode that rests on three assumptions. First, critical emotional praxis is grounded in a historical and political understanding of the role of affect in trauma culture and its implications for education. Second, critical emotional praxis consists in the ability to question affectively charged, cherished beliefs, exposing how privileged positions and comfort zones inform the ways in which educators and students recognize what and how they have been taught to see/act (or not to see/act) and empowering different ways of being with/for the Other.[1] Finally, critical emotional praxis translates these affective understandings into relationships, practices, and enactments that benefit teaching and learning for peace, mutual understanding, and reconciliation. This work invites educators and students to bear witness to trauma, suffering, and injustice, and it also prepares them to take responsibility for preventing future conflicts that perpetuate trauma. The affective implications of trauma in education encourage all those involved in schooling to understand their own preferences, values, and experiences and use those critically to initiate and sustain critical transformations in their lives.[2]

Trauma studies in the social sciences and the humanities originated in the context of research about the Holocaust, but they did not arise in a vacuum.[3] Trauma and its impact had been investigated by Freud, who showed that the ways in which one reacts to a traumatic event depends on his or her personal psychic history, that is, his or her memories and fantasies. Trauma—particularly *historical trauma,* the focus of interest in this book—refers to unthinkable catastrophic events that, when witnessed, evoke painful feelings. Psychoanalytic theories built on Freud's work treated trauma as an *unclaimed* individual experience that needed to be somehow dealt with, using various therapeutic strategies (Caruth, 1996). However, work in the post–Holocaust era has begun to engage trauma in a way that respects and contributes to its politics. Although trauma enters the social and political terrain as an expression of personal experience, explains Bennett (2005), "it is always vulnerable to appropriation, to reduction, and to mimicry" (p. 6); thus, the idea that trauma is "owned" by someone is deeply contested. In light of studies on the cultural politics of emotion, the concept of trauma has been redefined to include how traumas work to shape individual and collective bodies (Ahmed, 2004). As Kansteiner (2004) has even argued, trauma after the Holocaust has risen "as one of the key interpretive categories of contemporary politics and culture" (p. 193).

This book extends the notion of the politics of trauma in education by offering an account of how educators and students become invested in social and political norms that reify others as figures of evil and threat. Trauma, in this sense, is conceptualized as having an affective *force* that is both political and subjective; it is never unproblematically only "subjective" or merely "political"; rather, it is always lived and negotiated at an intersection (Bennett, 2005). The larger flows of power relations within which certain affective investments are made to social imaginaries (e.g., the nation-state, a religion, cultural heritage) forge particular subjectivities. If trauma testimonials entering educational discourses and practices are vulnerable to appropriation by ideological-political forces, an understanding of affective investments, I suggest, is fundamental for educators.

Feminist, critical, and poststructuralist thinking shows us that the politics of affects and emotions matters in many aspects of social life, including education (Zembylas, 2007); affects and emotions show us how power relations shape inclusion and exclusion boundaries between bodies—who should be *inside* and who should be *outside*. Exploring the restrictive and the productive forms of pedagogical engagement with trauma narratives does not offer us a privileged view of individuals, entire cultures, and nations; rather, by creating pedagogical spaces for critically exposing the affective politics of trauma, educators and students are invited to become aware of different modes of inhabitation. Alternative forms of embodiment, then, are enabled and invigorated by the capacity for transformation. The register of affective politics of trauma indicates the larger investments that are made and the ways that subjectivities are constituted—through affective investments such as hatred, shame, sentimentality, anger, fear, empathy, and so on.

This book, then, does not aim at simply describing the affective implications of trauma in education but at showing the way in which politics and ethics operate via the coding of affective investments and create openings for the production of new subjectivities. To follow Deleuze, affect or emotion is a catalyst for profound thinking and critical inquiry because of the way it engages us, "the impressions which force us to look, encounters which force us to interpret, expressions which force us to think" (1972, p. 161). Unlike some of the proponents of trauma studies, who are primarily concerned with the primary experience of trauma and its negative implications for individuals, the focus of this book is on the possibilities that are opened for individual and social transformation, as a result of interrogating the politics of trauma in educational contexts.

An important distinction that needs to be made here is one's positioning and context of encountering trauma (Kaplan, 2005). At one end, there is the direct trauma victim, while at the other, there is the person who has no personal connection to the victim but may encounter trauma through other sources such as the media and oral or written accounts of a catastrophe. It is in this context that the notion of *witnessing*—primary and secondary witnessing—has emerged in the field of trauma studies (Oliver, 2001). Becoming a primary witness to traumatic events is painful, because the witnessing experience shatters one's worldview, or as Felman and Laub write, the witness "becomes radically transformed by the very process of witnessing" (1992, p. 10). Oliver clarifies that there is a double meaning of witnessing: "*eyewitness* testimony based on first-hand knowledge, on the one hand, and *bearing witness* to something beyond recognition that can't be seen, on the other" (2001, p. 16, author's emphasis). For Oliver, it is the tension between eyewitness testimony and bearing witness that necessitates the infinite response-ability of subjectivity and therefore of ethical relations with the Other. Given that most educators and students encounter trauma through the media or literature, it is important to explore how they can engage in witnessing without falling into the trap of voyeurism/sensationalism or in melodramatic attempts to close the wound, as in some Hollywood treatments of historical trauma (see Kaplan, 2005).

Given the risks of re-presenting trauma and the challenges in forming empathetic connections between witness and testifier, listener and speaker, one wonders what forms of pedagogical engagement may be constituted to deal with trauma and what the limits of teaching and learning about trauma testimonies are. In this regard, Bennett (2005) suggests that it is better to understand the expression of trauma within a *transactive* mode rather than a *communicative* one. As she writes:

> [Trauma] often touches us, but it does not necessarily communicate the "secret" of personal experience. To understand its transactive nature, we need to examine how affect is produced . . . and how it might be experienced by an audience. . . . But if this affective transaction does not in and of itself convey the "meaning" of trauma, we must also pursue the question of how it might lead us toward a conceptual engagement. (p. 7)

Bennett's analysis refers to trauma-related art, yet it is relevant to other works that register lived experience and present it to an audience. This affective transaction is also evident in LaCapra's (2001) concept of *empathic unsettlement,* that is, the simultaneous feeling for another while becoming aware of a distinction between one's own perceptions and the experiences of the other.

What is generally lacking in the popular identifications with the victims of a traumatic event—especially in light of the September 11 attacks—is a conceptualization of critical affect that might enable educators and students not simply to feel for those victims and their loved ones but also to understand how trauma operates through affective connection and articulates its differences from other places around the world. Moreover, I propose that it is this context that provides a compelling basis for a pedagogical exploration of trauma and affect. A conceptualization of the pedagogical possibilities in view of these issues enables new directions for a radical reinterpretation of trauma and its implications, individually and socially. In this way, this project makes a significant contribution toward reconsidering the implications of hatred, shame, sentimentality, anger, fear, empathy, and reconciliation in contemporary pedagogical encounters and develops capacities for critical emotional praxis, which is a new level of ethical responsibility in education. This conjunction of affect, praxis, and criticality may be understood to constitute the basis not just of an empathy grounded on a feeling for others but also of a critical understanding of the structures of injustice and how injustices take place in different parts of the world. "Once this happens," writes Kaplan (2005), "one may feel obligated to take responsibility for specific injustices" (p. 23).

In what follows, I offer a personal account of experiencing the war of 1974 in my homeland and illustrate the complexity of the trauma—along with its political, ethical, and educational implications—as registered through my own experiences, growing up in a conflict-ridden area. Several events over the years altered my interpretation of trauma in Cyprus, the role of education, and my relationship with my homeland and its people and produced a new personal identity. In my account, the difficulty of juxtaposing the trauma of "my" community (the Greek Cypriots) with that of the other community (the Turkish Cypriots) emerges; one can see how education, the media, politicians, and other societal mechanisms contributed to dominant images and discourses about victims and perpetrators. I became aware of this only after I left Cyprus and spent several years studying in the United States. It is there, away from the toxic images and discourses with which I had grown up that I was able to engage in a different interpretation of the politics of trauma in Cyprus, the interconnections between individual and national trauma, and the hopeful openings for a different "translation" of past traumatic events through education.

It is these openings that I agonize to create and highlight through my work in this book.

Education in the Aftermath of the War of 1974 in Cyprus[4]

I grew up in a small village near the Troodos mountains, 1 hour west of the capital, Nicosia. After the war of 1974, this village, like many others in the area, became a refuge for families who fled from the north to save their lives. The Turkish military intervention divided the island, so many thousands of Greek Cypriots in the north and Turkish Cypriots in the south were displaced. Despite the fact that I was only 5 years old, I vividly remember many family friends and relatives, who became refugees, staying in our house for several weeks after those tragic events. I also recall how everyone was wearing black, mourning for what had happened. The only exception was my grandfather, a refugee from Morphou, who was always optimistic and was constantly saying that it was just a matter of days before the refugees would return to their homes: "Within fifteen days . . ." he would reassure everyone, "within fifteen days, and we'll be back home in Morphou." He kept repeating these words to friends and relatives, almost like a ritual, until he died in 1991 without ever being able to return to his beloved city—something that I had the opportunity to do in an emotionally overwhelming visit, after the opening of the Green Line in 2003.

I went to elementary school in the fall of 1975. One of the first childhood paintings I drew depicted the Turkish planes bombing Cyprus and the Turks as monster-like animals who wanted to eat us. This painting was put on display on a board and everyone reiterated how evil and barbarian the Turks were. I also remember participating in frequent commemorations of past historical events in which the Greek glories were celebrated. For example, we used to memorize all the heroes of the Greek revolution in 1821, and in our childhood games, each one of us picked a revolutionary hero and tried to "be" him or her. A few years after the war of 1974, the theme of DEN XECHNO (I don't forget) became prominent in our school life. Pictures of Kerynia, Bellapais, and Famagusta, our occupied places, would decorate all classrooms; the goal was to acquire knowledge so that we would never forget these places and care enough so that one day we would be ready to fight for them, if necessary. The most prominent themes of the DEN XECHNO campaign, which later became DEN XECHNO KAI AGONIZOMAI (I don't forget and

I struggle), focused on the remembrance of the Turkish invasion, the thousands of refugees, the missing, the enclaved, the violation of human rights, and the destruction of ancient Greek archaeological places. I recall how I was encouraged by a teacher to write letters to the missing persons (6 years after 1974) telling them how much we loved them and prayed for their return. All these images were part of the traumatic symptom evident in the media, the newspapers, the school textbooks, and our everyday social life. My parents told me many years later that during the first years after the war of 1974 I had often seen nightmares—planes were flying over Cyprus and bombing us, and everything was bursting into flames . . .

My educators presented the Greeks and the Turks in stereotypical ways: the Greeks as heroic figures who were always fighting for what was right, for justice, democracy, and freedom, and the Turks as barbarians, unjust, deceitful, evil, and war-loving. We were repeatedly reminded of what the Turks had done to us and that the young generation had a duty to remember and fight, if needed, to throw the Turks out of Cyprus. The perception in my mind about the history of 1974 was very clear: The victims who suffered were the Greek Cypriots, and the perpetrators who committed barbarisms were the Turks (in those years, I never made a distinction between Turkish Cypriots and Turks). Not a single teacher in my entire primary and secondary education discussed with us who the Turkish Cypriots were, whether they also suffered at the hands of Greek Cypriots, or whether Greek Cypriots and Turkish Cypriots lived in peace in the past and fought for common social issues (I only found out about these much later). On one of the few occasions in high school, when the issue arose of who the Turkish Cypriots were, I stood up in front of the class and repeated full of pride what I had been taught in Katichitiko (religious school): The Turkish Cypriots used to be Christians who became Muslims to avoid taxation during the Ottoman rule; the Turkish settlers from Anatolia never belonged to Cyprus . . .

As a university student studying to become an elementary school teacher, I participated in many anti-occupation marches organized by the Students' Union of the Pedagogical Academy and other organizations. Those marches were usually organized to honor past heroes and protest against the Turkish occupation of Cyprus. Occasionally, we were involved in "hot" incidents in which we confronted the Turkish troops in the Green Line, throwing stones at them or getting arrested by the UN Force for illegal entrance in the dead zone. I remember feeling proud when in one of those incidents, we were waving the Greek flag and yelling at the Turkish soldiers, who were almost face to face with us, "Cyprus is Greek!" In the evening, we were on the news; many friends and family members congratulated us for being "true Greeks," "fearless" and "bold" against those who occupied our land.

Several years later (in 1993), I left Cyprus to study in the United States. One day, while I was eating at the university's cafeteria, I was told by an American friend that there was another Cypriot in the cafeteria "but he doesn't have a Greek name." I was curious to meet this new Cypriot student, because I was certain that I knew all the Cypriots studying at my university. So, it was in the United States that I met the first Turkish Cypriot in my life . . . A few minutes after the initial shock of meeting each other, while many friends were watching us, we started yelling at each other: "You did (so and so) to us in 1974," I said; "You did (so and so) to us in 1963," he hit back. Then I replied, "Yes, but you did (so and so) to us in 1453" and he responded with another date that marked a traumatic event for his community . . . The conversation became so heated that the university security was called upon to intervene. We separated with a lot of lingering anger and resentment for each other, but I remember feeling proud that I had told him how evil "they" were. Fortunately, I met this Turkish Cypriot student many times after this troubling event, and we eventually became close friends. For many months, we had discussions along the lines of "You did so and so to us," and although there was no immediate shift in our perspectives about the history of Cyprus, we discovered that at least we could talk to each other in a civilized manner and hear each other's point of view.

I extended my studies at another university in the United States for several more years and at the same time continued to exchange e-mails with my Turkish-Cypriot friend. My studies in educational philosophy and theory taught me about the implications of socialization within nationalistic educational systems and the practices that had been used all over the world to instill hatred for the Other. It gradually became clear to me how national ideology worked to shape the ways in which the traumatic events of the war of 1974 were to be perceived by members of each community in Cyprus. Only then did I begin to discover the impact of all the dichotomies that were constructed in my mind as a young child: the "good" Greeks versus the "bad" Turks; "we," the victims versus "them," the perpetrators; and so on. The struggle to overcome these polarities was not easy but emotionally painful. I became angry with myself for being deceived for so long. I also felt much discomfort because I did not know how to feel about the new knowledge I had discovered. Confusion and ambivalence accompanied me for a long time . . . It was not easy to shatter the beliefs and emotions that I had carried with me for most of my life. But it was only then that I realized how I had never been taught to listen to the Other's point of view. I did not really appreciate my Turkish-Cypriot friend's perspective and all the discussions we had had until that moment.

After finishing my studies (in 2001), I returned to Cyprus and began teaching at a tertiary institution. I saw firsthand that my young students (high-school graduates) felt the same hatred as the one I had experienced several years before. My own personal transformation led me to make efforts to help my students rethink the way they had been educated: How dominant perspectives of memory and forgetting, trauma and suffering, were constructed and how toxic their consequences were. So now I teach my students about whether we sometimes need to learn how to forget in order to remember, so that we may gain a deeper understanding of our affective investments to sacred histories and face with courage what may be designated by others as "unpatriotic." My students' testimonies need to be heard; but the Turkish-Cypriots' testimonies also need to be heard and so I invite Turkish-Cypriot colleagues to come and present their views in front of a Greek-Cypriot audience. These testimonies intend not only to inform us about traumatic events and their haunting legacies, but also fundamentally to challenge us to alter our relation to these events and our modes of social interaction with each other.

Some time after the partial lift of restrictions on movement across the Green Line, I decided to visit my grandfather's house in the occupied territories. I was curious to see whether I could find the place (I was only 5 years old in 1974); what it would look like after all these years; who was living in the house; how I would feel, after carrying with me so many memories, photographs, and stories over the years. Would the house ceiling be as tall as I remembered? Would the mirror be hanging from the same spot on the wall? Would I find the small toy plane that I had forgotten on the kitchen table a few days before my grandparents had to flee for their lives? Silly questions, perhaps . . . The decision to pay this visit though was quite emotional not only because of the anticipation to see the "other part" of my homeland for the first time, but also because my family was adamantly against it. "Only when our country is free from those who occupy our land, we will go there," most of my family members kept emphasizing, when I announced my decision and asked if anyone wanted to go with me. When they realized how much I insisted, they threw at me their last remaining weapon: "You're a traitor, then! You're not a true patriot."

I left without saying much, and the next day I took on my mission to find my grandfather's house.

I drove through the village, parked the car, and walked toward the direction where I thought the house was. The anticipation mounted inside me. At the sight of the house, many childhood memories came back to me. To everyone's surprise (when I later narrated the story), I did find the house—without any explicit instructions by anyone and without the help of a map—just following my instinct . . . I knocked at the door and introduced myself. The family living there—Turkish-Cypriot refugees from the south of Cyprus—was naturally surprised to see me but then quickly welcomed me into the house. Their son (around the same age as I was) and I spoke in English, as nobody spoke each other's language. I told him how I had grown up in the south after the war, and he told me how he had grown up in that house during that same period. He said he had found a toy plane and had kept it all these years when the owners of the house would return, but then eventually he gave it to his 3-year-old son. I smiled. The toy was passed on to a child, just as it was supposed to . . .

I did not want to take any pictures of the house as I felt that it was an intrusion. I saw the house; I visited my childhood memories; and it was time to move on, I thought . . . "Next time, you should bring your family," they told me. The father took me in his arms and kissed me on both cheeks. "We are refugees too, you know," he whispered in broken Greek. "We want to go back to our own house in the south, and you should come and stay here. This is not our house. But you have to realize that we had to stay somewhere . . ." I felt bad; I did not want him to be apologetic. I left, unable to form any words on my lips . . .

In the summer of 2005, the Turkish-Cypriot friend from the university years and I met once again at a bi-communal social event held in a hotel on the buffer zone. I had not seen him since our university years, almost 10 years earlier. We both confessed to each other that an amazing transformation had taken place in our lives and that we were no longer the youthful nationalists we had been in the past! We agreed that memories of past traumas inflicted by one community against the other should not be dismissed, but it was time to move on. Move on where, we could not tell; moving on, we both admitted, should not be interpreted as forgetting but rather as a way of building connections between us. "Who would have ever thought that we would end up in the same position, advocating peace and reconciliation in Cyprus?" we said laughing.

The unexpected rupture in my personal life taught me the importance of *translating* trauma, that is, of finding new ways to make meaning out of traumatic experiences of the past. As Kaplan (2005) says, trauma can never be "healed" in the sense of a return to how things were before a traumatic event took place, or before one witnessed such an event; but the impact of trauma can be worked through individually and collectively. And here is exactly where critical emotional praxis can be an appropriate form of pedagogical engagement with trauma imagery. This book then moves through a focus on different manifestations of emotions and affects related to trauma—for example, hatred (chapter 3), fear (chapter 4), shame (chapter 5), and anger (chapter 7)—and argues that the impact of trauma can be usefully translated across pedagogical forms of critical emotional inquiry. This conjunction of affect, praxis, and criticality offers pedagogical opportunities of transforming perceptions for and actions with others.

The ethical and political importance of critical emotional praxis as an affective event and a pedagogical strategy is the degree to which, by cultivating a commitment to bear witness to trauma, it facilitates the emergence of a pedagogical space of affective potential. That is, to see the process of working through trauma as an act of witnessing involves negotiating affective connections that do not constitute us "voyeurs of the suffering of others, tourists amid their landscapes of anguish" (Ignatieff, 1997, p. 10). The call for critical emotional praxis in the classroom is affirmative and need not result in narcissistic, voyeuristic, or melancholic rhetoric; rather, the trauma can be worked through reconfiguring witnessing in affective, pedagogical, and critical terms.

CHAPTERS AND THEMES

Chapter 1 begins with an interrogation of the sentimentality, resentment, or desensitization in education as a result of the politics of emotion in the circulation of trauma narratives. Such an interrogation advises a different analysis of trauma narratives, one that acknowledges the politics of trauma and the dangers from its rhetoric. Sentimental education takes its strength by framing normative emotional responses to *all* traumas; it is grounded on empty empathy and unfounded optimism while giving the illusion of a just response. The tension between claims for universality or particularity of trauma requires a different pedagogical treatment of trauma narratives—one that moves beyond such a duality. It is argued that this tension needs to be critically interrogated in schools, and educators and students need to consider what it would take to subvert sentimental responses to trauma narratives.

Chapter 2 analyzes how the politics of trauma perpetuate conflict through escalations in fear, resentment, and hatred and thus pose a great challenge for educators who want to cultivate empathy and reconciliation in their curriculum and pedagogy. This chapter begins by examining two major theoretical strands in the literature on trauma studies—the psychology and the politics of trauma narratives—and using this analysis to highlight the role of collective memory and identity in relation to the politics of trauma. The politics of trauma is defined as an attempt to represent certain historical events in such a way that these events obtain a desired significance in the collective imaginary. Thus, the politics of trauma is about how past events are remembered and acquire hegemonic power and what challenges are created for educators in their efforts to critically evaluate narratives about historical trauma and suffering. This chapter also discusses how it might be useful to consider the contribution of the notions of empathy and reconciliation to make educators and students aware that they are falling prey to the hegemonic power of trauma narratives. It is argued that these concepts prove helpful in addressing the concerns raised here, providing a fruitful avenue for curriculum inquiry and pedagogy.

Chapters 3, 4, 5, and 7 take on four particular emotions—*hatred*, *fear*, *shame*, and *anger*—that work as important affective orientations in the politics of trauma in education. In chapter 3, I show how the notion of *emotional ambivalence* can be helpful to educators who struggle to address the pedagogical implications of students' feelings of ethnic hatred. I suggest that although hate feelings are difficult to change, unraveling the ambivalence in the affective politics of hatred creates openings for enriching educators' and students' perspectives on mutual understanding and solidarity with others. Drawing on insights from an action research project conducted in a higher education institution in Cyprus, I show how the notion of hatred as ambivalence constitutes an important step to initiate a subversive analysis.

In chapter 4, I explore the affective politics of fear in education and discuss the implications for educational policy, theory, and practice. The aim is to highlight the ways in which fear, in its unavoidable invasion to the pedagogic field, *educates* students to create particular affective relations to one's self and to others. In this manner, I argue that narratives of fear function as influential pedagogies of emotion, and as such they are particularly effective ways of placing teachers and students within a conflicting ethos of life. Analyzing the implications of the affective politics of fear in the context of education is important for educators who want to establish a different affective community in the classroom, a community that is not paralyzed by fear and trauma but that deconstructs arbitrary boundary formations and encourages an ethos of mutual understanding and reconciliation.

Against a perspective of shame that would see it as something paralyzing, negative, and destructive, chapter 5 draws out a different account of the politics of shame in education. I explore the openings that are created for a *productive* politics of shame in the context of discussions surrounding intercultural education via tracing how different kinds of shame are evoked or rejected in collective imagination—both in politics and in education. My goal is to show that a rehabilitation of the political and ethical value of shame challenges the opposition between self and Other and that this act can be beneficial, particularly in educational settings. An effort to transcend the usual dichotomies around the pride/shame debate both in political and in educational discourses will tackle the cultural, political, and educational uses of shame and help educators gain a deeper understanding of the meaning and implications of shame in discussions about trauma.

Chapter 6 (written with Zvi Bekerman) builds on the background ideas explored in chapters 3, 4, and 5 and discusses whether justice can be addressed in a world of uncertainty and multiplicity of identity that produces powerful emotions such as hatred, fear, and anger, as a result of defining one's identity through the negation of the Other. This chapter also asks how the dilemmas of justice can create useful openings for the ethical and the affective—not only for the epistemological—implications of peace and coexistence education. These questions are concerned not only with how to actively engage children to *know* how to be just in an uncertain world, but also with how to *inhabit* affective spaces that seriously consider the affective relations with the Other in a manner that is vigilant to the emotionality of justice. The purpose of this chapter, then, is to turn our attention to the ways in which the emphasis on the pragmatics of justice is implicated in the politics of trauma in education and particularly to the affective politics of pedagogies that promote peace building and coexistence.

In responding to the hegemonic politics of trauma in education, chapter 7 suggests that the politicization of anger is not only inevitable but also desirable. In particular, an analysis of anger in education may offer certain critical advantages, including a better understanding of the importance of *being angry* in political terms. As various educational approaches nowadays seek indiscriminately to silence anger in the name of "emotional intelligence"—and thus, de-politicizing it—an attention to the political aspects of emotions in education is extremely valuable. Thus, this chapter argues that anger is central to the exercise of power relations in the school, and it challenges the dominance of instrumental/rational accounts about anger replacing them with the recognition of anger as political. Also, some of the conditions that generate angry feelings as ambivalence are explored in light of the transformative possibilities that this ambivalence creates in the context of trauma testimonies.

Chapter 8 (written with Zvi Bekerman) culminates many issues raised in previous chapters through exploring the meanings and implications of *dangerous memories* in two different sites of past traumatic memories: one in Israel and the other in Cyprus. Dangerous memories are defined as those memories that are disruptive to the status quo, that is, the hegemonic culture of strengthening and perpetuating existing group-based identities. The effort here is to outline some insights from this endeavor—insights that may help educators recognize the potential of dangerous memories to ease pain and offer hope. Central to the analysis of the implications of the politics of trauma in education in this chapter is the narration of two stories from longitudinal ethnographic studies on trauma and memory in Israel and Cyprus; these stories are interpreted through the lens of dangerous memories and their workings in relation to the hegemonic powers that aim to sustain collective memories. The two different stories suggest that collective memories of historical trauma are not simply "transmitted" in any simple way down the generations—although there are powerful workings that support this transmission. Rather, there seems to be much ambivalence in the workings of memories that under some circumstances may create openings for new identities. The final section discusses the possibilities of developing a pedagogy of dangerous memories by highlighting educational implications that focus on the notion of creating new solidarities without forgetting past traumas. This last section employs dangerous memories as a manifestation of critical emotional praxis in the context of my overall concerns about the implications of memory and trauma for the education of individuals and communities in conflict-ridden areas.

Finally, in the afterword, I consider how the various pedagogical engagements proposed throughout the book constitute forms of *aporetic mourning* in educational theory and practice. I argue that the work of mourning can evoke public and school pedagogies that provide an alternative way of relating to otherness and trauma—not through remaining fixated on simply re-presenting the other's or one's own trauma, but rather through the insistence on remaining inconsolable before suffering. A major concern addressed in several chapters of this book has been the normalization of trauma in school and public discourses through the establishment of boundaries between *grievable* and *ungrievable* lives. The violence unleashed through national mourning and the reproduction of loss through rituals of commemoration in schools raise important issues about the ethics and politics of mourning embedded in public and school pedagogies. Therefore, I summarize the arguments put forward throughout the book by arguing that the recognition of our vulnerability to loss, as the groundwork to the formulation of school and public pedagogies of aporetic mourning, can induce a powerful point of departure for political alignment with others; this alignment has the potential to move us beyond trauma's seduction.

CRITICAL EMOTIONAL PRAXIS[5]

Running through the chapters of this book is a certain notion of critical emotional praxis that foregrounds the pedagogical engagement with the politics of trauma. Acknowledging the politics of trauma in education compels educators and students to interrogate their affective investments as part of a larger project to build empathy and reconciliation with the Other. Critical emotional praxis, then, denotes the struggle of transgressive educators to create pedagogical spaces in which those affective investments can be challenged. In the chapters that follow, it is shown that the constellation of *difficult emotions* related to traumatic experiences is hard to avoid in the classroom. Sometimes experienced consciously, at other times on a less than fully conscious level, these emotions are socially and politically relevant, rather than deriving solely from one's individual idiosyncrasy. By drawing on these emotions and enacting critical emotional reflexivity, educators and students can engage in interrupting normal(ized) educational practices. This critical emotional praxis can serve as the basis for educational work on the politics of trauma. This point of engaging and reflecting on the politics of trauma in education is precisely where a study of affect, praxis, and criticality advances educators' understandings of how to create spaces for interrogating unjust, conflicting, and exclusionary educational practices (see also Giroux, 2004a, b; McLaren, 2003).

First, critical emotional praxis begins with the acknowledgment that educators and students need to develop an understanding of the role of affect in trauma culture and its impact on them and on their respective communities and cultures. They need to examine the emotional implications of unjust, conflicting, and exclusionary systems and practices, that is, how emotions work to make various categorizations that include some individuals or groups and exclude others

(Ahmed, 2004). Structures of conflict, violence, and hatred depend upon withholding particular emotional responses (such as grief, remorse, compassion, and caring) toward certain groups of people deemed as *others*. Schooling would indeed be well served if educators and students created opportunities to engage in intrapersonal reflection on their emotional understanding of historical trauma-related issues. Critical emotional praxis calls on educators and students to unpack their cherished beliefs and comfort zones in order to deconstruct the ways in which they have learned to see, feel, and act. They must recognize in what ways emotions resulting from traumatic experiences define how and what one chooses to see, and conversely, not to see (e.g., how nationalist dynamics perpetuate hatred and conflict, and how empathy for the others' suffering often involves discomfort for one's own identity). Critical emotional praxis for trauma in education, therefore, builds on the work of those who have argued that affect needs to be politicized in education in order to address questions of otherness, difference, and power (Boler, 1999; Boler & Zembylas, 2003; Zembylas, 2005, 2007; Zembylas & Boler, 2002).

The aim, then, of a teacher who is willing to enact critical emotional praxis is to analyze the *emotional landmines* (Boler, 2001, p. 1) created in the classroom community, in order to understand and challenge how they shape and mark one's sense of emotional attachment and identity. This process is emotionally discomforting and unsettling for both students and educators, but under some circumstances, discomforting learning helps educators to engage in new affective relations with others. Developing the skills and knowledge to analyze how educational practices teach people to feel the world through an ideological lens, often with little awareness they are doing so, is an important step in identifying trauma, fear, hatred, and anger.

Finally, critical emotional praxis draws from the above emotional understandings and responds to the particular context in which a school is located, creating and enacting teaching and learning practices that reconceptualize the emotional culture of a classroom/school. For example, others' trauma cannot be shared simply through empathy; neither the creation of a compassionate culture in the classroom deals with the bigger issues that significantly shape students' and educators' emotional responses. However, research shows that, under some circumstances, a teacher's persistent and creative efforts to enrich the emotional culture of a classroom/school can indeed alter the emotional connections between students and teachers, even in the face of larger issues of school culture (Berlak, 2004; Zembylas, 2005, 2007). These connections are not made as a form of fellow feeling, and they are not about feeling the other's trauma but about the possibilities of rethinking our relations to trauma (Ahmed, 2004). For educators and their students, this implies learning to live with the dangers of a knowledge about others' trauma that can interfere with its own sentimentalization.

In this context, critical emotional praxis can open windows of opportunity for counternormative work that offers "possibilities of interruption" (Weis & Fine, 2001, p. 521). Given the issues we all know that await educators and students, especially in conflict-ridden areas, it is important to pay more attention to developing, analyzing, and enacting strategies for navigating through the emotional challenges of traumatic experiences so they can become sites of positive change, not negative impairment. In other words, critical emotional praxis can serve as an important tool for educators and students struggling to enact pedagogies that increase the possibilities of interrupting the prevalent politics of trauma.

What seems to be missing in the current conceptualization of trauma testimonies in education is not a consideration of how educators and students can feel for the victims of a traumatic event but to articulate and understand their difference from those victims. How, for example, Greek-Cypriot victims are different (but also in some ways similar) from Turkish-Cypriot victims; how witnessing the September 11 attacks differs from witnessing in the West Bank or in Johannesburg. We need to consider how educators and students make sense of themselves, their practices, and their political options in relation to their emotional understandings of trauma and the ways such understandings affect their actions—both locally and globally. For the most part, the issues discussed in the chapters of this book are the product of a sustained engagement with the consequences of the politics of trauma in education and the possibilities of interrupting this politics; these chapters do not offer any didactic political statement. Insofar as they warrant a pedagogical engagement, they do so because they point to different ways of *(un)doing* the politics of trauma in education through an aesthetic of *relationality* with others. The politics of trauma in education is enacted within a sphere of relationality, and as such, within this same sphere the constitution of new subjectivities and new alliances has to be pursued and sustained. These alliances are needed to change the *affective ethos* of schools; it is a transformation of the often-pervasive conflicting ethos through the invention of alternative educational spaces that might enable renewed coalitions and solidarity between individuals and across communities. Yet it should be clear that these educational spaces do not guarantee any revolutionary transformations; educational interventions by themselves cannot go far without institutional changes (Bekerman, 2007b; Bekerman & Maoz, 2005; see also chapters 6 and 8).

Enacting relationality means, for example, sharing stories of loss and trauma within critical educational spaces. Juxtaposing different stories of loss and vulnerability, without, however, effacing the specificity of each loss or making losses equivalent, provides a critical space to problematize hegemonic identity categorizations. The call for questioning the dominant politics of identity is not a call for erasing identity by universalizing it on the basis of suffering, because this idea would replace one hegemonic position for another (Bekerman, 2007a). Along these lines, students need to become knowledgable not only of their community's trauma stories; when other trauma stories are heard and analyzed, the students begin to deconstruct the positions premised on the existence of clearly differentiated identities based on race, ethnicity, or culture. New manifestations of "us" emerge from engaging with the other's suffering, and a different learning experience begins when the affective space is provided to emphasize the historically constructed nature of identity and culture (Appiah, 1994). This process—which is long and emotionally painful (Zembylas, 2007)—could be liberating, argues Bekerman (2007a), because "freeing identity from its tight historical political connections to the development of nation-state could reinstate identity as an enriching difference" (p. 33).

Allowing for the recognition of what identity and culture *do* in everyday life, through acknowledging the other's suffering and grief, evokes counternormative educational spaces. In these spaces, educators and students can begin to study stories of trauma and mourning from conflicting communities (Metz, 1972). It is within this context that educational spaces can become unsettling, critical, and hopeful, while propelling individuals and groups into a new process of affective alliances and solidarities that may subvert existing identity categories. Reclaiming forgotten connections with others demands acts of compassion, self-criticality, and resistance to homogeneity and normativity. Engaging with trauma is indeed not an easy task; yet it is a noble and courageous task, one for which the contributions of education are vital.

Chapter 1

The Violence of Sentimentality in Encounters with Traumatized Others

How does trauma enter education? Do students' and teachers' emotional responses to trauma constitute moral principles against violence and injustice? How do feelings of injustice morph the rhetoric of trauma narratives that are circulated through curriculum and teaching? Trauma has often been described as a private, psychological experience, as a painful event that an individual goes through but that others cannot feel or fully understand (Herman, 1997; Kaplan, 2005). And yet trauma is continually evoked in public and pedagogical discourses as that which demands collective response (Ahmed, 2004). Deemed otherwise unspeakable, the unrepresentable trauma is translated into narratives and testimonials that are effective in communicating trauma as a painful experience (Berlant, 2001, 2004). Thus, it is argued that introducing students to trauma narratives and testimonies is an important way of approaching controversial issues such as war, cross-cultural conflict, genocide, racism, and terrorism (Bouchard, 2002; Schweber, 2004; Zembylas, 2006b, 2007). For example, feelings of sadness or moral anger about others' suffering enable students to enter into a relationship with others: "the pain of others becomes 'ours,'" writes Ahmed, "an appropriation that transforms and perhaps even neutralizes their pain into our sadness" (2004, p. 21). The first chapter of this book is grounded on the notion that teaching about stories of suffering, pain, and trauma allows for learning that is emotionally, politically, and educationally significant; yet, it is also pointed out that in this process there are several dangers that need to be critically analyzed and dealt with.

On the one hand, Ahmed (2004) asserts that having feelings about others' trauma aligns us with them. The emotional distress or inspiration that accompanies this alignment grounds the claim for our possible political transformation and our critical commitment to keep the traumatic event open, which means communicating it in a way that keeps it traumatic for others (Berlant, 2001). On the other hand, however, as Berlant (2001, 2004) warns us, this alignment does not imply any identification with someone else's trauma, and it may lead to the trauma being fetishized, which condenses the trauma in self-repetition and habituated banality. That is, the argument about the particularity of trauma and the impossibility of identifying with someone else's suffering not only leads to an impasse in terms of connecting with others but also threatens to diminish the implications of trauma.

In other words, mobilizing either the universality or the particularity of trauma and suffering does not necessarily lead to any transformation. Bombarding students constantly and unwisely about collective traumas in the world can circulate a continuous and displeasing negativity. In fact, I see three equally dangerous student responses to educators' attempts at registering the inhumanity of violence through the use of trauma testimonials in schools: first, a sentimental reaction by students who identify with privilege and respond defensively yet feel uncomfortable and guilty, fearing that they will be exposed as immoral by refusing to bear any longer a population's collective suffering; second, an intense resentment by those who feel subordinated and may eventually get stuck in victim politics; and third, the desensitization of the student-spectators who get irritated by the scenes of suffering in some way, refuse engagement with it or minimize its effects, misread it conveniently, and reduce it to a few pedantic phrases.

In particular, there is a thin line beyond which the overvaluation of trauma narratives in schools can lead to what Berlant (2000) calls "national sentimentality"; that is, a liberal rhetoric that avows that a nation can feel the burden of others' trauma and suffering through channels of affective identification, compassion, and empathy (see also Berlant, 2004). Hence, the call to feel others' pain and suffering is key to the liberal narrative of compassion (Woodward, 2005). However, as Spelman (1997) argues, "Compassion, like other forms of caring, may also reinforce the very patterns of economic and political subordination responsible for such suffering" (p. 7). The overrepresentation and overvaluation of others' trauma fixes them as those who have trauma and who can overcome that trauma only when others feel compassion for their suffering. Traumatic experiences, which are in part effects of socioeconomic relations of violence and poverty, are assumed to be alleviated by the empathetic understanding and generosity that are enabled by such socioeconomic relations (Ahmed, 2004). The case of compassion and empathy based on such an assumption is not only ideological but also naïve (Woodward, 2005). As Woodward rightly asks, Under what circumstances are empathetic understanding and generosity translated into policies that alleviate suffering?

Clearly, I do not mean to suggest that trauma narratives and the ways they are handled in schools are invariably sentimentalized or do not contribute toward building solidarity. I am suggesting, however, that sentimentality is indeed a popular rhetorical means by which viewing trauma is advanced as a moralistic means of promoting good citizenship and moral character (Berlant, 2000). Consequently, the intensity with which memorial practices of trauma and violence are utilized in contemporary society—take, for example, the way many modern nation-states commemorate past victories over or defeats by an enemy—demands an examination of the politics of emotion promoted and sustained in education through the rhetorical use of trauma narratives and testimonials. Critical engagement with such narratives "is not a linear process but one of constant return and reassessment" (Scott, 2001, p. 359).

Undoubtedly, every educator faces unsettling challenges—rhetorical, emotional, political, and ethical ones—when he or she pays attention to the current proliferation of testimonies to violence and trauma. Notwithstanding the power of such testimonies to evoke solidarity, the questions that this chapter asks are uncomfortable: What happens when the testimonial form becomes ordinary in our classrooms over time, operating in sentimentalized packages that eventually produce so much misrecognition of the socioeconomic realities in the world? What does it mean for educators to shape students' collective response when a politics of sentimentality threatens to shape the discussion and analysis of trauma narratives in the classroom? Finally, what happens to issues of dealing with alterity when feeling bad about others' trauma becomes evidence for injustice and feeling good becomes evidence of justice? The discomfort revealed in these questions will not be easily overcome by ignoring the politics of emotion embedded in the publicized traumas of subordinated populations. Educators need to consider what it would mean to induce "a break with trauma's seduction" (Berlant, 2001, p. 36) in their pedagogies. First, though, I must trace the theoretical connection between emotion, trauma, and claims of justice that grounds this discussion.

THE POLITICS OF EMOTION AND CLAIMS OF JUSTICE

How exactly do emotions work to "show" the effects of injustice while generating openings for healing and recovery? Or as Ahmed puts it: "Is a just response to injustice about having more 'just emotions,' or is justice never 'just' about emotions?" (2004, p. 191) Ahmed explains that there is a difficult relation between emotion and (in)justice; to transform bad feelings into good ones is not necessarily to repair the costs of injustice. In other words, the meaning of injustice cannot simply be reduced to one's feelings about injustice, because this reduction would be deeply problematic at two levels (Berlant, 2000). First, such a reduction would essentially make an individual's judgment of what is right or wrong dependent upon the existence of specific emotions; that is, if bad feelings are present, then and only then can injustice be identified. Second, this position would claim access to the interiority of others' feelings, a claim that is impossible.

Contemporary emotion theory reveals that moral principles and justice claims are not established based on universal emotions; instead, emotions are indicators of moral beliefs—they do not constitute them (Solomon, 1990). Consequently, feelings of disgust and anger about the violence and trauma inflicted on others are indicators of our beliefs about justice, but justice itself has to be established first (Karstedt, 2002). In fact, as Karstedt explains, there is longitudinal research on the development of very young children that shows that they understand moral principles but do not attach moral sentiments to them until a later stage, around the age of 10. In other words, the cultural diversity of emotions shows that emotions are linked to antecedent moral judgments but do not constitute them.

Nonetheless, it has to be stressed that feelings for others and their suffering provide signs of obligation to others. Thus, injustice takes the form of perpetuating particular kinds of affective relations to social norms that are responsible for people's traumatic experiences (Ahmed, 2004). Challenging injustice, then, implies the interruption of these social norms and consequently the subversion of the affective relations to them. Although injustice cannot be measured by the existence of suffering, as Ahmed emphasizes, some suffering is an effect of injustice. But if injustice is not simply about feeling bad, then justice is not simply about feeling good. The important idea here is that one way of understanding injustice is to describe it as the failure to connect with others and respond to their suffering. Needless to say, this is not all that injustice is; however, this thesis emphasizes a usually neglected aspect—the connection between social norms and the affective economies that maintain such norms (Zembylas, 2007). It is not uncommon, then, that feelings of fear, hatred, and resentment are constructed as a result of social and political struggles around justice and identity claims.

Consequently, it seems important for educators to help students understand that justice is not simply a matter of feeling empathy for those who suffer. For instance, feeling solidarity for poor children in Africa is not an indication of justice in the world. When some individuals are constructed as "grievable" (Ahmed, 2004, p. 192) to us—as opposed to others who may be deemed "ungrievable"—it is often through the sentimentalization of loss, something that is deeply troubling (see also Butler, 2004). Similarly, feeling bad about others' trauma does not necessarily suggest that injustice is removed. For example, does mourning for children's injuries as a result of the millions of landmines that are spread in conflict-ridden areas around the world repair the injustice of the injury? Is our judgment about what makes a terrorist act bad dependent upon the other's suffering or our disgust for such an act? Again, these questions are not meant to provide a superficial description of injustice; they simply expose the complex relationship between feelings and injustice.

The analysis so far offers two important perspectives in building my argument about the politics of emotion and its relation to justice claims. First, any attention to emotion needs to be placed within a particular social and cultural context if it is to be an effective force in understanding the ethics and politics of justice claims. In other words, this idea suggests that emotions—especially those that are associated with trauma, such as fear, anger, and shock—are perceived as ethical and political technologies belonging to a situation, as opposed to organic emotions that "grow" out of individuals (Thrift, 2005). For example, emotions that are associated with trauma narratives on the Holocaust make sense if they are understood not only as individual but also as collective responses to specific sociopolitical conditions, both in the past and in the present.

Second, the politics of emotion does not preclude a productive and creative sense of living that engages with the world through affective interventions rather than through "private bargains with misery" (Thrift, 2004, p. 66). In other words, a politics of emotion does not only involve risks for educators who want to teach about trauma, but also can forge a renewed criticality that involves a critical mobilization of emotion in order to produce new understandings of social relations and new ways of being in the world (Zembylas, 2007). The search for this criticality constitutes the centerpiece of this book.

THE POLITICS OF TRAUMA, MEMORY, AND RHETORIC

The term *trauma* may be defined as the response to an overwhelming event that cannot be fully assimilated by those who experience it at the time it happens, and the trauma victims continue to find it difficult to come to terms with its extreme emotions such as terror, fear, and shock (Burns, Busby & Sawchuk, 1999; Caruth, 1995, 1996; Farrell, 1998; LaCapra, 2001; Leys, 2000). "The trauma," writes Caruth, "is a repeated suffering of the event, but it is also a continual leaving of its site" (1995, p. 10). Under normal circumstances, painful memories of the past are often forgotten (forgetting does not imply amnesia here); however, "the mind reconstructs its memories under the pressure of society" (Halbwachs, 1992, p. 51). For example, it has been repeatedly argued by social and political scientists that nation-states utilize various rhetoric mechanisms to instill particular events (as either glorious or traumatic) in the collective memory of citizens (Bodnar, 1992; Fentress & Wickham, 1992; Winter & Sivan, 1999). Although many historians criticized Halbwachs's analysis on collective memory (see Kansteiner, 2002, 2005), an interesting tension that arises out of his position is the interrelatedness between past and present, on the one hand, and individual and social, on the other. To put it differently, there is a tension between personal grief and collective capacity to feel—a tension that does not seem to settle easily (Bennett, 2005; Berlant, 2004; Kaplan, 2005).

On the one hand, it may be argued that the impossibility of feeling the pain of others does not mean that the pain is, strictly speaking, theirs; an ethics and politics of responding to trauma involves its sociality (Ahmed, 2004). In other words, traumatic experiences are deeply affected by encounters with others in the society. On the other hand, this position does not imply that pain is universally the same; on the contrary, the politics of emotion emphasizes that the notion that trauma is not a private or an interior state but the formation of power relations and the operation of socioeconomic norms (see Foucault, 1990a, b, c) is embedded in the constitution of traumatic experiences. This view is inspired by the writings of Foucault, who interrogates the discursive and disciplinary places from which questions of self are posed and insists that it is impossible to claim any originary and socially uncontaminated self that is coherent and unified. In Foucault's writings, the unified self—who strictly owns trauma, for example—is challenged; as he shows, there are manifold ways in which individuals are historically constituted.

Thus, it is important to acknowledge that the phenomenon of trauma takes place in a dialectical space between individuality and sociality (Svašek, 2005). Unavoidably, then, traumatic experiences are not only remembered as initially experienced but are also remembered by how they are "perceived by others, directly or indirectly, and the extend to which others are able to listen empathetically to a survivor's testimony" (Brison, 1999, p. 42). That is, trauma stories create a social and political space in which victims reconstruct memory and may heal or perpetuate their pain depending on how others respond to their trauma. A nation-state may co-opt the personal grief of individuals to perpetuate feelings of anger and resentment against a rival group. After all, it is common practice for nationalist politicians to select and incorporate particular historical narratives and memories of trauma into their political rhetoric to evoke and strengthen collective narratives (Cairns & Hewstone, 2002; Svašek, 2002). As Ramanathapillai (2006) emphasizes: "The tone, emphasis, passion, and energy of a political narrative can be used to rally and unite people with a single goal. But the danger of political narratives of trauma is that they are vulnerable to manipulation and can create and perpetuate power for unscrupulous leaders" (p. 5). Even when no direct loss is incurred, argues Fierke (2004), individual identity and collective emotion are bound up in the politics of trauma: "The emotions may remain disguised in individuals, but, to be translated into political agency and identity, they must be put into words by leaders, who give meaning to the individual experience by situating it in a larger context of group identity" (p. 484).

Similarly, critical theorists emphasize that schooling is not a neutral ground but a place in which the dominant culture's ideology (e.g., how to teach certain trauma narratives) is imposed (Apple, 1979; Giroux, 1997; McLaren, 1989). Donald (1992) in his book *Sentimental Education* takes up Foucault's analysis and critically examines education as "a political way of maintaining or modifying the appropriation of discourses, along with the knowledges and powers which they carry" (p. 46). The performative power of schooling is manifested in the cultural norms that are embodied in what and how things are taught. As Donald argues, there is a dynamic interconnection "between the pedagogic and the performative." This analysis can be applied to the analysis of trauma narratives to show the role of education (as well as that of media and popular culture) in the "formation of a national culture" (p. 48) as well as "national subjects," to use Bhabha's (1990, p. 297) term. In other words, individuals and collectivities come to be constructed as victims or perpetrators through a range of national discourses.

Consequently, there are clearly rhetorical, political, and ethical implications of narrating stories about trauma, even though the stories can never actually repeat or represent what happened (Caruth, 1995, 1996). Thus, it is one thing to experience trauma and quite another to construct or hear stories about it. If there is an uncomfortable sense of colonization and a subsuming of the experience of the Other within rhetoric means, it is because this danger is always present. This position does not diminish the meaning and implications of pain and suffering. On the contrary, Spivak (1987) insists that those who are recipients of stories such as trauma narratives have a tremendous responsibility: first, not to presume that suffering can be understood universally, and second, to be vigilant about misuses of such stories. There is never anything transparent or universal about the meaning of *wound*, which means that knowledge about the wound may become a property of rhetoric (Berlant, 2000, 2001).

It is precisely within this space that reconstructed memories and emotions can often be translated into a sentimental rhetoric. The sentimental framing of suffering, Berlant argues, is problematic because it wrongly assumes such suffering as universal. The pain of slavery, for example, cannot be fully understood by a white, middle-class reader of a book on slavery; the politics of personal feeling cannot address the institutional aspects of injustice (Woodward, 2005). The experience of being moved by sentimental scenes of suffering, insists Woodward, may work instead to return us to a private world far removed from the public sphere. Thus, failing to address the structural reasons for injustice and simply using personal stories to tell about structural effects risks subsuming the very attempt to understand trauma narratives into a passive ideal of empathy.

For this purpose, Brown (1995) warns us that there is a contemporary fetishization of the wound in subaltern politics. She asserts that there is an obsession with memory as a symptom of late modernity, arguing that the "wounded attachment" (p. 71) to traumatized identity—that is, the investment of subordinated subjects in the wound—comes to stand for identity itself. For Brown, the politics of memory essentially becomes a politics of trauma involving claims of injury. However, as she asserts, the political claims of injury are inadequate as a basis of politics and hold no promise of redemption, because such claims make action impossible. Brown insists that any political desire attached to redeeming the past simply "reinfects the narcissistic wound" (p. 72) and destroys the desire for change. Ahmed (2004) reiterates that the transformation of the wound into an identity is problematic because of its fetishism. She writes: "The transformation of the wound into an identity cuts the wound off from a history of 'getting hurt' or injured. It turns the wound into something that simply 'is' rather than something that has happened in time and space" (p. 32).

I agree that the transformation of the wound into an identity is problematic, but a politically conservative response to this kind of argument may conclude that whatever happened in the past belongs to the past. Therefore, in a crippling irony, Ahmed's argument about historicizing the wound may backfire. Consequently, I suggest that a possible response to this tension should not be limited to a duality of either remembering or forgetting past wounds; there are clearly numerous other issues involved in the process of how this happens, and individual/collective uptake of trauma narratives is only one aspect of this (see Kaplan, 2005). The critique of the politics of wounds should not operate as generalized critique—something that clearly amounts to a strict and stern paternalism (Woodward, 2005). So a good response to this tension would part company with claims that our task is simply (not) to forget the past. Instead, in order to move beyond the rhetorical use of wounds, we need to explore the different ways in which suffering is politicized.

Clearly, not all narratives of trauma are the same; Ahmed is right to warn us that the media and other social institutions (e.g., schooling) often turn trauma into a form of spectacle, in which the pain of others is commodified. There is a market for suffering that appropriates trauma in sentimental ways. While Ahmed touches only superficially on the dangers from this commodification of suffering (Kleinman, Das & Lock, 1997), Mestrovic (1997) moves a step further and attends to the development of blasé attitudes (Simmel, 1971)—habitual tendencies to dismiss the seriousness of trauma to protect one's self from the constant bombardment of the intense trauma narratives that are circulated.

Mestrovic (1997), in fact, has proposed the concept of "postemotionalism" to describe this kind of blasé response to trauma narratives, arguing that "contemporary Western societies are entering a new phase of development in which synthetic, quasi-emotions become the basis for widespread manipulation by self, others, and the culture industry as a whole" (p. xi). The emotionalization of public discourse about trauma and justice is a response to changes in wider emotional culture and the moral imagination of Western societies (Barbalet, 1998; Boltanski, 1999; Williams, 2001). The "postemotional type," in Mestrovic's terminology, takes cues from peers, the media, educational institutions, and cyberspace as to when one "should rationally choose to exhibit a vicarious indignation, niceness or other pre-packaged emotions," yet he or she is "unable to put these emotions into appropriate action" (p. xii).

In a postemotional society, a "new hybrid of intellectualized, mechanical, mass produced emotions has appeared on the world scene" (Mestrovic, 1997, p. 26) as a response to narratives of collective suffering. The media engage the public in distant suffering (Boltanski, 1999)—compassion and sympathy for the victims and expressions of moral disgust toward the perpetrators (Karstedt, 2002). The "McDonaldization of emotions" represents the "pre-packaged, rationally manufactured emotions—a 'happy meal' of emotions—that are consumed by the masses" (Mestrovic, 1997, p. xi) whenever a story of trauma is circulated. This attitude occurs when individuals become allergic to involvement yet intelligent enough to know that events are significant.

Here there are, then, three equally important dangers associated with responses to trauma narratives. First, the audiences become highly emotionalized moral spectators in the distant suffering of victims and perpetrators (Boltanski, 1999). This reaction begins and ends in the emergence of sentimentalism—especially by those who are privileged—that is mediated by guilt and pity. Such a reaction, argues Mestrovic (1997), will degenerate into a new form of totalization that is so nice, tolerant, and charming, whatever the event, that it proves hard to resist. This form of politics of emotion transforms sentimentalism into expertise over the meaning of feelings of (in)justice. That is, certain modes of thinking and feeling about the performance of (sentimental) emotions become expert techniques regulated by social norms grounded in claims to truthful knowledge (see Rose, 1990, 1998, 1999).

Second, another danger comes from the response of those who feel victimized and entails a feeling of resentment. Based on Nietzsche's notion or *ressentiment,* Meltzer and Musolf (2002) explain that resentment comprises feelings of hatred, wrath, revenge, and the like. This desire for revenge, as Brown (1995) has shown, may indeed be a reaction against unjust treatment, but attachment to past wounds refuses to see the future possibilities of action. "Revenge as a 'reaction,' a substitute for the capacity to act," as she writes, "produces identity as both bound to the history that produced it and as a reproach to the present which embodies that history" (p. 73). For Brown, there is a paralyzing tendency inflecting the logic of wound attachment; a preoccupation with the past risks forcing subaltern subjects to get stuck in a present with no hope and a future that puts the past before everything. Butler (1997a), on the other hand, provides an alternative view and argues that the trauma of hate-speech brings about opportunities for agency through the constitution of a subject who can retort (see chapter 7). In other words, getting stuck in the weight of history is not necessarily the only option; the circulation of trauma narratives may open up forms of subjecthood from which the subaltern can exercise agency. This direction may indeed rise to the condition of possibility beyond both cynical reason and an empty commodified sentimentalism based on falsely shared suffering.

Finally, there is the danger that the sentimental narrative is consumable and ineffective (Woodward, 2005). The moral spectators of others' suffering are irritated by the scenes they see; thus, somehow they become unwilling to engage with the implications of suffering and injustice. An unforgiving position would be to argue that this desensitization is morally bankrupt, just as Berlant (2000) claims. But I prefer to argue that desensitization is dangerous as a political position because my concern falls less on feelings-as-moral and more on involvement-as-political. Being desensitized essentially means being unwilling to do something about others' suffering.

Let me rephrase what I have been arguing thus far in terms of the rhetoric of trauma narratives and the dangers from their unwise (pedagogical) use, that is, the use of trauma narratives in ways that do not attend to the tensions and the complexities described here and lead students instead to an uncritical sentimentality. In its simplest sense, this chapter asks, What does it mean for testimonies of trauma to become a rhetorical form that must be told endlessly while their authority is undermined by the sentimental or resentful feelings infecting the politics of trauma? Central to the emergence of the politics of trauma is the fetishization of the wound and the refusal to construct a political language with which to address the radical openness to the future (Kilby, 2002). Instead, the logic of circulating trauma narratives, especially in conflict-ridden areas, has been to reiterate the weight of history. Yet there are serious consequences from this preoccupation with traumatic stories that must be told endlessly: "the numbing or banalizing aspects of repetition," as Berlant (2001, p. 47) writes, threaten to sentimentalize and eventually erode the truths of testimonies. I find such critique compelling, especially when it is shown that these repetitions are "ones that pierce a will to no longer feel vital or alive, and therefore traumatized" (ibid.). However, as the earlier references to Butler's work (1997a, 2004) suggest, Berlant and Ahmed's critiques are not the only viable options; Butler (see also 1997b) provides a different perspective by arguing that acts that render us vulnerable may also create the conditions for interruption. In fact, in the last part of this chapter, my effort focuses on sketching some ways through which educators may oppose the banality of compassion and the paralyzing resentment that spread an ideology of sentimentalism in education.

THE VIOLENCE OF SENTIMENTALITY IN EDUCATION

Textbooks, the media, and school curricula play an important role in cultivating trauma narratives and the rhetoric around them (Cole, 2007; Salomon & Nevo, 2002). Consequently, the pervasiveness of collective narratives of trauma, collective historical memories, and their transgenerational transmission through schooling make students and teachers particularly susceptible to the influence of the rhetoric of trauma. How could educators, then, respond to the danger of trauma narratives being fetishized? The dangers described earlier do not necessarily shed light on who decides what and when something becomes fetishized. However, a trauma narrative becomes fetishized when it becomes commodified and fails to provide new ways that help us gain a deeper understanding of others' suffering.

Following Ahmed's (2004) account, one could suggest that a response to the fetishization of trauma narratives should attend to the different ways in which trauma enters educational discourses. Such a move implies that not all trauma narratives should be read in the same manner—for instance, as forms of entitlement. As Ahmed asserts, to read the story of white male injury in the same manner as the stories of subaltern injury would be an unjust reading. In other words, power relations in specific social and political settings cannot be ignored; subjectification has to be understood in a complex of apparatuses, practices, machinations, and assemblages within which human beings are fabricated (Rose, 1998). This notion also implies that reading and critiquing trauma narratives in school settings cannot be done in a generalized manner; attention needs to be paid to the different ways in which the politics of trauma shape the rhetorical forms of the narratives.

However, Ahmed's claim may inadvertently lead to privileging one person's trauma over another. There are two major issues with this claim. First, who has the right to choose between the validity and weight or trauma—especially in school settings, given the debates over the ideological mission of schools? Second, reverse racism accusations could be made; in other words, one cannot assume that differences between white male injury and subaltern injury are either essential or incidental in any way. The *subaltern* as well as the *white male* as categories should not be reified uncritically. These issues bring to the surface once again the tension between the universality and particularity of trauma—a tension that is not resolved by choosing one or the other. In order to break the duality with which this tension is posed, a response from educators could be the problematization of taken-for-granted boundaries between categories in search for ways that witness the heterogeneity of trauma narratives. I would further suggest that students and educators engage in critical interrogation of issues that are defined as polar opposites—for example, forgetting/remembering, universal/particular, public/private; reading trauma narratives needs to involve reconsideration of existing hierarchies and essentialized categories.

Central to the emergence of acknowledging and critiquing the politics of trauma has been the work of scholars and activists from within feminist, antiracist, and postcolonial studies. In the field of education, for example, it is argued that painful experiences require a radical reconfiguration of the way emotions are understood (e.g., see Boler, 1999; Boler & Zembylas, 2003; Zembylas, 2005, 2006b, 2007; Zembylas & Boler, 2002). Acknowledging that trauma becomes a property of rhetoric does not diminish trauma, regardless of where it comes from. However, it is important for educators and students to oppose the sentimentality that pursues collective cohesion by circulating the notion that trauma is universal—that is, the idea that trauma is universally the same. Trauma has a history and a politics, and educators and students cannot afford to ignore these elements. Unavoidably, then, trauma narratives have to be analyzed in the realm of political action. At the same time, the history and politics of trauma should be critically analyzed in ways that do not perpetuate hegemonic historical categorizations.

It is important, therefore, to reiterate that stories of trauma must indeed be heard in schools; to think otherwise is to be shallow and insensitive. But the conditions of possibility for hearing them must also be interrogated because empty empathy (Kaplan, 2005) threatens to slide through the doors. Trauma is not an emblem of true wisdom about injustice; however, as it has been argued, hearing trauma narratives in schools without sentimentalizing them is a daunting task. The challenge for educators and their students is how to learn to hear what is painful. I am afraid that such hearing is not possible if trauma is claimed as either one's own exclusively or universally the same. It is one thing to claim that testimonies cannot be taken away from others; however, it is a completely different thing, a step forward in my view, to learn to hear the other's pain and respond to this pain as witnesses and not as spectators (Simon, 2005; Simon, Rosenberg & Eppert, 2000).

Many privileged students, argues Wang (2005), resort to rational arguments or sentimental reactions and fail to acknowledge how their own emotional attachments affect the way they know. Laden by the denial of guilt, they resist that they are part of the history of trauma. Some students respond to trauma narratives and the discomfort involved "by choosing to reinforce their own identities rather than risk self-transformation, while others feel overwhelmed and depressed by the dark side of history and culture, from which they have been sheltered" (Wang, 2005, p. 58). In all of these responses, the desire to use trauma narratives as a means of showing the pain of subordination can easily lead to emotions of pity for those who suffer or feelings of resentment on the part of those who constantly feel subordinated (Chizhik & Chizhik, 2002).

Also, the desensitization potential of using trauma narratives is a particularly important effect that needs to be considered. It is possible that the fetishization of trauma in media and literature will disengage students from the discomforting implications of traumatic events (Boler, 1999). The spectating student may initially feel disturbed by frequently televised images of a traumatic event, but fatigue from the repeating scenes of suffering fixes the event in a few images and produces a decontextualized view on injustice. As Kaplan (2005) writes, the visuality of suffering in catastrophic events—for example, September 11, 2001, in the United States—is often translated in such melodramatic or sensationalized ways that these events hardly seem real. The result is that students reduce the meaning of such events in a few superficial or exaggerated phrases (Boler & Zembylas, 2003; Zembylas, 2006b).

To shatter the strong grip of sentimentality, resentment or desensitization is essential to develop pedagogies that acknowledge the discomfort caused by trauma narratives and transforms such feelings into energy for praxis and transformation. Martusewicz (2001) urges educators and students to engage in what she calls a *pedagogy of suffering*, while Boler (1999) refers to it as a *pedagogy of discomfort* (see also Boler & Zembylas, 2003; Zembylas, 2007). These are examples of pedagogies that deal with the inner emotional battles of students who struggle to reach new ground. Brown (1995) goes one step further and suggests that the desired transformation of individuals who read trauma narratives can only happen if trauma is treated as ideology, not as knowledge that causes discomfort. In my previous work (Zembylas, 2007), I have discussed various possibilities of critical pedagogies that problematize teachers' and students' emotional attachments to particular ideological commitments and suggest the creation of pedagogical spaces of critical hope. Thus, it is possible to develop *pedagogies of critical hope,* that is, sites of intersubjective encounters that entail transformative possibilities. In the following chapters, I extend this discussion and offer a number of possibilities that emerge from an engagement with the politics of trauma in education.

Fighting against the violence of sentimentality, resentment or desensitization in education is crucial, if teaching and learning about/from trauma narratives in schools are not going to end up making these narratives banal stories. Using such narratives to teach students that the good life will be achieved when there is no more trauma in the world while the structures of sentimental normativity are kept intact will indeed make banality unavoidable. Such a realization advises a different analysis of trauma narratives, one that acknowledges the politics of trauma and the dangers from its rhetoric. Sentimental education takes its strength by framing normative emotional responses to all traumas; it is grounded on empty empathy and unfounded optimism, while giving the illusion of a just response.

A strict emphasis on the particularity of trauma will not remove the danger of banal sentimental responses to trauma narratives, and neither will an emphasis on trauma's supposed universality. Empathy is not a bad strategy, as I show more explicitly in the next chapter, but it requires a future that is not drowned in a "sentimental contract" (Berlant, 2000, p. 44) with past trauma. The cost of this contract, writes Berlant, "is the muffling of an analytically powerful and political rage, [and] an equivocation of demand and radical critique" (p. 44). This is why it is imperative for educators and their students to place at risk the sense of good feeling that sentimentality promises—not because the bonds of empathy "are false or socially harmful, but because they so frequently make ethical a refusal to counter the customary and structural violences of social life, which are deemed somehow extraneous in the face of pain's claims" (Berlant, 2000, p. 45). In order to move beyond the dangers of sentimentality, resentment, and desensitization, trauma narratives should constitute a pedagogical ground in which renewed political vocabularies of justice must be fought for. That is, educators' and students' task is to constantly reinvent new ways of becoming critical witnesses of trauma narratives.

Chapter 2

The Politics of Trauma in Education

In Erich Marie Remarque's *All Quiet on the Western Front,* Paul, the main character, a German soldier during World War I, talks to a soldier he has stabbed and held in his arms until his death.

> Comrade, I did not want to kill you . . . But you were only an idea to me before, an abstraction that lived in my mind and called forth its appropriate response. It was that abstraction I stabbed. But now, for the first time, I see you are a man like me. I thought of your hand-grenades, of your bayonet, of your rifle; now I see your wife and your face and our fellowship. Forgive me, comrade. We always see it too late. Why do they never tell us that you are poor devils like us, that your mothers are just as anxious as ours, and that we have the same fear of death, and the same dying and the same agony—Forgive me, comrade; how could you be my enemy? (1982, p. 223)

Paul's words echo the destructive ethos of conflict: dehumanization of the opponent; the *us* and *them* mentality; the systematic cultivation of hatred for *the enemy.* Dehumanization, the process by which people are viewed as less than human, is a process accompanied by a wide range of negative emotions toward them, such as contempt, hatred, or fear. The readers can feel Paul's unequivocal regret for not accepting the Other as human until it is too late. In this way, Paul maps the process of *rehumanization* (Halpern & Weinstein, 2004), when he *empathizes* with the enemy and finally sees him in human terms. The major function of empathy is imagining the particular perspective of the Other—that is, realizing that the Other is like me, you, and him. Finding commonality through identification with the enemy is perhaps the most difficult and yet profound step in the rehumanization of the Other.

The emotional effects of group conflict and trauma on individuals and large groups and the prospects of healing are significant concerns for educators in conflict-ridden societies (Danesh, 2006; Salomon & Nevo, 2002). Conflicts deeply involve society members and result in the construction of a *conflictive ethos* (Bar-Tal, 2000) that provides the dominant affective orientation to the society. This affective ethos of conflict includes particular perceptions, emotions, beliefs, and attitudes—for example, pride about the in-group, hatred about the adversary group, and unwillingness for any intergroup relations—all of which must change for *reconciliation* to occur (Kriesberg, 1998; Lederach, 1997; Rothstein, 1999). Reconciliation refers to a societal-cultural process in which new emotions and beliefs that encompass respect, coexistence, and peace are formed about an adversary (Bar-Tal, 2000). The ideal of reconciliation in conjunction with the emotional effects of conflict and trauma raises a question that has been largely ignored in the education literature. Can (and should) educators work to contribute toward the goal of reconciliation when conflict and trauma have a social and political manifestation that is larger than the sum of traumatized students and teachers? If so, how is this possible—politically, pedagogically, and emotionally?

This chapter analyzes how the politics of trauma perpetuate conflict through escalations in fear, resentment, and hatred and thus pose a great challenge for educators who want to cultivate empathy and reconciliation in their curriculum and pedagogy. In general, the theme of politics of emotions emphasizes that emotions are not private or interior states, but they are crucial to the formation of power relations and the operation of social norms in everyday life contexts (Lupton, 1998; Lutz & Abu-Lughod, 1990). This does not mean that emotions do not involve subjects but rather that what characterizes emotions is the lack of individual residence and their involvement in relations of power (Ahmed, 2004). For example, as shown by various scholars, the work that emphasizes the *individual* perspectives of trauma-related emotions (e.g., fear, resentment, and hatred) has disregarded the *collective* dimensions of those emotions (Biton & Salomon, 2006; Fierke, 2002; Svašek, 2005; Volkan, 1997; White, 2000). The collective memory of fear, hatred, victimization, and dehumanization becomes a powerful symbol and an effective tool to strengthen the existing conflicting ethos (Fisher, 1997).

Consequently, I argue, when the emotional elements of the politics of trauma are not accounted for in peace education efforts, they risk perpetuating the existing conflicting ethos. In spite of work showing how educators can respond to conflict through practices of peace education and conflict resolution (Burns & Aspeslagh, 1996; Davies, 2004; Wintersteiner, Spajić-Vrkaš & Teutsch, 2003), there is surprisingly little investigation to understand the effects of politics of trauma and its emotional consequences in relation to the role of education. It is in this sense that it is argued that understanding those effects and building empathy and reconciliation may provide openings to make peace and coexistence a long-term option that challenges the cycle of conflictive ethos (Bekerman & McGlynn, 2007; McGlynn et al., 2004; Shechter & Salomon, 2005). At the same time, it is important to emphasize that such educational interventions in and of themselves are not a substitute for structural change (Bekerman, 2007b).

In the first part of the chapter, I examine two major theoretical strands in the literature on trauma studies and use this analysis to highlight the role of collective memory and identity in relation to the politics of trauma. The politics of trauma may be defined as an attempt to represent certain historical events in such a way that these events obtain a desired significance in the collective imaginary. For example, proclaimed trauma victims—both survivors of trauma and people who claim that they have been indirectly affected—use discourses and practices of collective victimhood in an attempt to gain political influence (Scheff, 1994; Volkan, 1997). Thus, the politics of trauma is about how past events are remembered and acquire hegemonic power and what challenges are created for educators in their efforts to critically evaluate narratives about historical trauma and suffering. In the next part of the chapter, I discuss how it might be useful to consider the contribution of the notions of empathy and reconciliation to make educators and students aware that they are falling prey to the hegemonic power of trauma narratives. I argue that these concepts prove helpful in addressing the concerns raised here, providing a fruitful avenue for curriculum inquiry and pedagogy.

THE PSYCHOLOGY AND POLITICS OF TRAUMA NARRATIVES

Generally speaking, victims of historical trauma (i.e., war, genocide, and mass atrocities) suffer physical, mental, and emotional pain, because they find it difficult to come to terms with extremely shocking events (Caruth, 1995; Leys, 2000). In current literature on trauma studies, there are two major theoretical strands of analysis that strive to build and develop the consequences of trauma and ways of healing trauma in our lives. In this chapter, I refer to these two strands as the *psychological strand*—which includes psychiatry and medicine—and the *political strand*, respectively—which emphasizes the discourses in which trauma narratives are politicized. The purpose of this section is to review what each strand has to offer and highlight the understanding of trauma through the lens of emotion and affect. This lens will provide the ground for appreciating the contribution of the notions of empathy and reconciliation in dealing with the politics of trauma in education; these notions have not only deep affective roots but also important political overtones.

The Psychological Strand: Memory and Identity

The psychological literature focuses on the individual element of trauma, that is, what is happening within an individual (Fierke, 2004). Medical and psychiatric studies have pathologized trauma through discourses of hysteria and posttraumatic stress disorder and focused on the abnormality of trauma (Kleinman & Kleinman, 1991). The notion of the detached individual who is traumatized has been reinforced by dichotomous theories of identity that have made clear distinctions between individual/society and private/public (Cohen, 1997). Yet, from an experiential perspective, emphasizes Svašek (2005), self/other distinctions are always blurred because identifications with ethnic (or other) groups are central to self-identity.

Given that self-identification is inextricably intertwined with a large-group identity, the study of traumas is a key link between individual and large-group psychology. This link becomes more complex for traumas arising from intergroup conflicts (compared to those from natural disasters, for instance), because the pain in such cases is a result of intentional action of one group against another (Zinner & Williams, 1999). Genocide, for example, is a paradigmatic case of intentional act that results in collective trauma. The intentionality argument partly explains the role of collective trauma in the construction of a shared identity (Tajfel & Turner, 1986, 2001), collective historical memories, and the politics of trauma. The enemy's intentions to cause horrible acts of violence connect the individual to his or her large group with a profound sense of belonging: Individual identity is subsumed to the group (Petonito, 2000).

A different perspective that moves beyond the focus on the individual or the evaluation of trauma as abnormal is found in the pathbreaking work of psychoanalyst Vamik Volkan (1988, 1997, 2001; Volkan & Itzkowitz, 1994). In his work, Volkan applies psychoanalytic theory to examine various ethnic, religious, and national conflicts, in relation to cultural identity and ethnic attachment, and analyzes how traumatic events are transmitted from one generation to another. He asserts that traumatic narratives are chosen to fit a coherent story of victimization and define a large-group identity; for this purpose, he uses the concept of *chosen trauma*. Volkan defines chosen trauma as "the collective memory of a calamity that once befell a group's ancestors. It is . . . a shared mental representation of the event, which includes realistic information, fantasized expectations, intense feelings, and defenses against unacceptable thoughts" (p. 48). Within virtually every large group, writes Volkan (2001), "there exists a shared mental representation of a traumatic past event during which the large group suffered loss and/or experienced helplessness, shame and humiliation in a conflict with another large group" (p. 87).[1] Similarly, his concept of *chosen glories* refers to the mental representations of a large group that include a shared feeling of success and triumph—for example, celebrations of independence day and ritualistic recollections of glorious events and national heroes in schools (see Bekerman & Maoz, 2005).

An ethnic group's chosen trauma defines its identity "by the transgenerational transmission of injured selves infused with the memory of the ancestors' trauma" (Volkan, 1997, p. 48). That is, hatred in one generation seem to "pass" to succeeding generations in the form of a "psychological DNA" (1997, p. 46). According to Volkan (2001), the transgenerational transmission is linked to the unresolved group mourning, that is, the group's "inability to mourn losses of people, land, or prestige," and indicates the "group's failure to reverse narcissistic injury and humiliation inflicted by another large group" (p. 87).[2] During these "communal memories of violence" as Apprey (1999) calls them—that is, commemorations of a lost battle or defeats of the enemy—the group experiences the chosen trauma (or chosen glory) as if the past were in the present, and members of the group react to current events with emotional intensity to reverse ancestral humiliation (or to feel proud in order to do the same as one's ancestors did). It is as if succeeding generations and especially the youth are assigned the task to avenge the honor of their ancestors (Weingarten, 2003). Thus, "the toxic errand of extinction, humiliation, massacre, a legacy of ashes" (Apprey, 1999, p. 135) is passed on to succeeding generations. The inability to mourn a chosen trauma and the evolution of defense mechanisms against it influence the social and political ideologies of a traumatized group and may result in a new generation embracing an ideology of revenge (Volkan & Itzkowitz, 1994). As Mack (in Volkan & Itzkowitz, 1994) writes:

> It is remarkable how little empathy is felt by national groups for the suffering of their traditional enemies, even if the victimization on the other side is palpably evident and comparable to or greater than one's own . . . The lack of empathy, the inability to identify with the anguish experienced by the members of a national group toward whom one bears hostile feelings, removes one of the central deterrents to the waging of war. (p. 10)

The result of this *egoism of victimization,* as Mack (1990) calls it, is a ceaseless "cycle of victimization" (p. 125). An example of this kind of discourse is found in Volkan's (1979) early work in which he used psychoanalytic theory to analyze the ethnic conflict between Greek Cypriots and Turkish Cypriots. In his study of the historical conflict between Greeks and Turks on the island of Cyprus, Volkan described several mechanisms that show how a child born in Cyprus would absorb a dominant large-group identity defined by national sentiments—that is, whether one is Greek or Turk—and not any other large-group identification (e.g., religion).[3]

The Political Strand: The Politicization of Trauma

In addition to seeing trauma as a psychological phenomenon, it is important to broaden its analysis and consider it as *political* too. This kind of understanding is promoted by sociocultural theories of emotions (Ahmed, 2004; Barbalet, 1998; Lupton, 1998) in which the expression and experience of emotions is grounded in social and political relations. Understanding trauma within a sociopolitical framework focuses on the operation of cultural and political discourses and on the ways traumatic emotions are produced in this context. This view, which is grounded on Foucault's work, enables the exploration of trauma as a productive rather than as a constraining force over individuals and groups. By focusing on the productive force of trauma rather than on the constraining conditions of an inner psychological reality, scholars can understand trauma as a form of *emotional politics* produced within particular national, religious, or political discourses. Thus, for example, trauma is not viewed as an isolated psychological event or as an ahistorical construct, but it exists *relationally* within particular historical conditions (e.g., nationalism, racism, religious fanaticism, sexism) that call attention to the nature of discursive practices that constitute feelings of victimization about one's own group and hatred for an adversary group (Card, 2002; Eisenstein, 1996; Scheff, 1994).

To speak of trauma, then, is to bring it into the political world; the question is how this politics is expressed and what the consequences are (Fierke, 2004), especially in educational settings. It is important to highlight that the acknowledgment of the politics of trauma points to the complexity of the connection between the individual and the large group, the personal grief and the collective human capacity to feel. The failure to recognize this connection and the resulting tension will increase the likelihood of not appreciating the powerful emotional forces of trauma at various levels. For example, teaching students of a traumatized group that they believe in the wrong cause (e.g., nationalism) is not a very encouraging strategy for conflict resolution (Kreuzer, 2002). This strategy will fail not only for psychological reasons but also because the rhetoric of trauma narratives has to be problematized.

Salomon (2004) points out that collective trauma narratives and collective historical memories constitute two major challenges that peace education faces and has to deal with. In the history of intractable conflicts, various groups have used traumatic narratives to perpetuate conflict and justify their actions based on the argument that their enemies' intentions are full of hatred and prejudice (Ramanathapillai, 2006). Textbooks, the media, and school curricula contribute to the cultivation of such narratives and their historical underpinnings (Davies, 2004). According to social identity theorists (Tajfel & Turner, 2001), mutual prejudice and the formation of a group identity tend to increase when perceived threat and fear emerge between conflict groups. As it is shown in the following chapters, the politics of hatred (chapter 3) and fear (chapter 4) make groups vulnerable and perpetuate collective narratives of trauma. The failure to recognize that the suffering of others is part of *our* goals and trauma will likely increase the probability of future suffering (Fierke, 2004).

In general, the psychological and political writings on trauma offer substantial insight into the emotional politics of trauma narratives and highlight two important issues that need to be seriously considered in educational efforts to reclaim trauma narratives in ways that promote empathy and reconciliation. First, in trauma-oriented groups, the individual's identity becomes more strongly linked to his or her group identity and enemy perception. The emotional relationship between the individual and the large group—how the individual, for instance, derives much of one's sense of self and self-worth from identification with one's national group—is something that needs to be seriously taken into consideration in any (peace) education program (Salomon & Nevo, 2002). At the same time, the prevalent conception of the Other as a fixed category—that is, the enemy—dangerously increases the tendency to dehumanize the Other. As Petonito (2000) points out, constructing people categories and subsuming individuals into a category makes people forget that human beings comprise it. Since the dehumanization of the Other is one of the ways that a group scaffolds its own sense of identity, education has to reverse this process by rehumanizing the Other (Salomon, 2004; Cole, 2007).

Second, the study of the *affective economies* (Ahmed, 2004) created by past historical events and their workings in the present (e.g., in a classroom) is important in grasping the ways in which children's emotional expressions to those events are gradually constructed and inflected. The term "affective economies" denotes that emotions do not reside in individuals, but they circulate in relationships of difference. Such an argument clearly challenges the assumption that (traumatic) emotions are individual and private phenomena and supports the position that emotions (and more generally affects) are political in the sense that power is an inextricable aspect of how bodies come together, move, and dwell. Affective economies may establish, assert, subvert, or reinforce power differentials, because affectivity separates us from Others as well as connects us to Others; this is why it functions as an economy. This economy, as Ahmed points out, is social, material, and psychic. Inevitably, then, an effect of this circulation is the constitution of emotional attachments and meanings, that is, emotions become attached to objects, bodies, and signs—a process that is crucial in the constitution of subjectivity (e.g., "national subjects"). Affective economies work to differentiate some bodies from other bodies—a process that is never completely resolved.

Consequently, educational efforts that challenge the politics of trauma (e.g., traumatic memories and nationalist sentiments) are complex, dynamic, and take a long time. For example, it is immensely challenging for peace education programs to assist young children to come to terms with the traumatic events of their group's history, to mourn but not to be consumed by the past, to accept the reality of a significant loss, and to change individuals' dispositions and actions toward the collective narratives of the Other (Danesh, 2006; Salomon & Nevo, 2002). In particular, mourning and healing may be used by educators to help children accept losses, subvert the egoism of victimization, and become able to face more realistically their disappointments, hopes, and aspirations. Therefore, educational programs for children should provide productive opportunities that identify and reevalute the heavy emotional load of trauma narratives—both one's own collective trauma narratives and the Other's narratives.

The following brief overview of the notions of empathy and reconciliation constitutes a conscious attempt to begin reclaiming trauma narratives in *productive* ways. The decision to outline some thinking on these notions so early in the book is clearly purposive and aims at setting the theoretical background of what will follow in the chapters to come; that is, my task here is to show not only the complexity in the notions of empathy and reconciliation but also their productive potential. Empathy and reconciliation are put forward as major critical components in the re-formation of educational goals in conflict-ridden societies; as such, these notions have enormous potential for educational programs, because they are grounded in redefining relations to traumatic events and thus reimagining relationships to/with the Other.

EDUCATION THAT BUILDS EMPATHY AND RECONCILIATION

First, how is education, especially the trauma narratives that are taught, related to the broad ideas of empathy and reconciliation? I argue that there are two important connections that need to be highlighted. The first is related to the idea that any process of reconciliation at the macro level (e.g., at the level of the state) cannot ignore the significant impact that micro-level reconciliatory processes may have, and schools are grounded closely enough to the lives of average citizens so that they can make useful contributions at this level (e.g., by reevaluating the emotional load of past narratives, as pointed out earlier). The second aspect is related to the critical intervention of educational institutions in problematizing the existing structures of injustice that put hurdles in efforts toward reconciliation and empathy. This aspect requires confrontation with the status quo and the mobilization of critical education and social justice pedagogies to address the issues that compromise any reform effort.

In building empathy and reconciliation, a wide variety of alternative narratives need to be developed out of the mutually hostile trauma stories. It is important to deepen awareness and criticality in children about how trauma stories can be used to teach fear, hate, and mistrust (Ramanathapillai, 2006) and establish a violence-based worldview (Danesh, 2006). All narratives, Kreuzer (2002) emphasizes, even the ones from the perpetrators of violence, need to be considered seriously, because they help us understand the emotional aspects of conflict and they point toward openings for strategic intervention. To build empathy and reconciliation in education programs, it is valuable to identify the narratives that evoke fear, hate, and mistrust and publicize the stories that show positive emotions emphasizing the humanity of the enemy—for example, stories of collaboration and caring among adversary groups. Telling positive stories can help rehumanize the Other and counteract the confrontational symbolical and emotional content of competing narratives that work to dehumanize the enemy. I suggest, therefore, that the promotion of empathy and reconciliation in education programs is a critical component of developing alternative narratives about past traumas—narratives that contribute to the transformation of the hegemonic conflictive ethos.

Why Empathy and Reconciliation?

Recent scholarship on the notions of empathy and reconciliation sees them in many different ways that pose complex and difficult demands and highlights the importance of rescuing those terms from the dangers of sentimentality (Cole, 2007). This work emphasizes how reconciliation at the macro level cannot be separated from efforts at the micro level. While some scholars are more interested in placing an emphasis on the role and meaning of reconciliation as a vehicle of wide social changes (e.g., Opotow, 2001), others focus on more practical terms and maintain that the essence of reconciliation is about bringing people together to relate and create a mutual understanding through dialogue (e.g., Lederach, 1997). Again, it is important to clarify that there is no dichotomy between these two moves; both ideas are grounded in the notion that reconciliation cannot be rigidly defined on universal standards, but it is a multifaceted and contingent process that has to take into consideration both the macro and micro levels.

Opotow (2001) writes:

> Reconciliation requires not only bringing people together to create a shared understanding but to succeed, much more. It requires an unflinching confrontation with the underlying, chronic injustices faced by a society and the mobilization of its institutions to address these issues in ways that are distributively and procedurally just, and genuinely *inclusive*. (p. 167, added emphasis)

Inclusion has become something of a cliché in recent times (Benjamin, 2002); yet, at this point I join others' efforts to expand the meaning(s) of inclusion (e.g., see Lingard & Mills, 2007) and provide links to issues of social justice and especially to reconciliation. In particular, seeking to establish connections among inclusion, social justice, and reconciliation requires serious work that includes new theorizations as well as novel state policies and educational practices. This task demands a willingness to see macro and micro levels as sites of *inclusive pedagogies* for reconciliation. The meaning of "pedagogies" is used here in the widest possible sense, that is, spaces and practices for learning (Shor, 1992) in which cross-cultural and sustainable discourses enable the "production, dissemination, and circulation of [critical] ideas" in the public sphere (Giroux, 2004b, p. 106).

Lederach (1997) explains that reconciliation is a social and political space, a "point of encounter . . . between the open expression of the painful past, on the one hand, and the search for the articulation of a long-term interdependent future, on the other hand" (p. 29). He identifies truth, mercy, justice, and peace as energies that enable reconciliation but acknowledges the tensions and paradoxes that exist between them. One of the tensions is that reconciliation involves both the individual and collective aspects. Needless to say, there is almost always an unbridgeable gap between collective and individual efforts for reconciliation. For example, in the case of collective reconciliation, a highly contentious issue is the question on behalf of whom one has the right to reconcile with the Other. Undoubtedly, collective reconciliation operates within larger ideological and political contexts. But, there is interdependence between the individual and collective aspects of the reconciliation process. Therefore, Lederach suggests that interventions at multiple levels of leadership (macro and micro ones) and within multiple systems are critical to achieving reconciliation.

Furthermore, Lederach emphasizes that relationships are crucial to the process of reconciliation. He emphasizes this in the following manner:

> The relationship is the basis of both the conflict and its long term solution . . . Reconciliation is not pursued by seeking innovative ways to disengage or minimize the conflicting groups' affiliations, but instead is built on mechanisms that engage the sides of a conflict with each other as humans-in-relationship. (1997, p. 26)

It is within relationships that empathy is developed for the Other's humanity, and the process of transforming polarized trauma narratives of the past is initiated. In this sense, empathy is an expression of *being in* relationship with the Other. To overcome the intensive negative feelings of trauma narratives—resentment, anger, hatred, and the desire for revenge—the door of empathy has to open; it is the development of empathy that reverses the process of dehumanization and returns humanity to the Other (Halpern & Weinstein, 2004).

Undoubtedly, empathy is a highly contested term that is employed by an unusually diverse group of educational thinkers, both from the left and the right (Verducci, 2000).[4] Theorists consider empathy from different ideological premises and thus define it in different ways and for different purposes. Consequently, empathy cannot be considered a unitary concept and thus, I do not claim that my analysis here claims any single framework as *the* empathy. I will provide an understanding of empathy that can be used to help identify its significance for education programs and pedagogies that wish to promote peace, reconciliation, and mutual understanding.

For the purposes of this book, then, empathy is the ability to incorporate other people's perceptions and *see* an experience with their eyes (Halpern, 2001). For example, a Serb youngster empathizes with a Bosnian Muslim when he is willing to see ethnic cleansing from the other's perspective; and a Greek-Cypriot student empathizes with a Turkish-Cypriot peer when she is able to see the Cyprus conflict from the other's viewpoint. Halpern explains that empathy is distinguished from sympathy in that the latter is about experiencing shared emotions with another person, while the former involves imagining and seeking to understand the perspective of another person. Both empathy and sympathy involve experiencing emotional resonance with someone else; however, this is not sufficient for empathy. Empathy goes a step further and combines a critical effort to imagine what someone is facing. According to Halpern, this imaginative element presupposes accepting the Other as a distinct individual and thus challenges the process of dehumanization in trauma narratives. Empathy does not make it impossible to dehumanize the Other; it just makes it harder, and that might just be a good thing. For example, it will be less likely for a Jewish student to dehumanize his or her Palestinian peer, if feelings of empathetic understanding are developed (Shechter & Salomon, 2005).

Halpern and Weinstein (2004) suggest that there are three important qualities in empathy that make it particularly important for reconciliation. First, it entails seeking the individual perspective of another; this idea can serve as an important motivation and encouragement to place one's self in another's circumstances and to accept that he or she possesses the same rights. Second, empathy involves a genuine effort to get to know the Other and his or her perspectives. In contrast, polarized trauma narratives close one's mind toward the

Other's experiences. Third, empathy involves emotional as well as cognitive openness and the tolerating of ambivalence; this ambivalence entails viewing differences as an enriching part of creating an ongoing workable relationship (Zembylas, 2007). For these reasons, becoming able to empathize with the Other's experiences and narratives is probably the most challenging goal for peace education (Salomon & Nevo, 2001).

This willingness to *feel with* the Other's perspective does not imply cognitive identification with the Other's views. In other words, the idea of accepting the Other's collective narrative as legitimate does not imply agreeing with it (Salomon & Nevo, 2001). Rather, empathy entails the *critical* and *analytical* ability

> to accept that the other side, in ways quite similar to one's own side, is so deeply motivated to fulfill underlying needs and express their own values fully, to overcome past traumas and insure future safety and expression of identity, that coercion or suppression of those needs and values will probably be counter productive. (Rothman, 1998, p. 225)

Emotional resonance is used to guide imagining into the individual experience of another person; this resonance is fundamental in beginning to construct shared meanings and evaluations (Halpern, 2001). In this way, the possibilities that a common basis of sameness is developed "by the sheer act of trying to see, interpret and feel as your opponent sees, interprets and feels are maximized" (Kreuzer, 2002, p. 9). Accordingly, as Staub emphasizes, this effort is instrumental in restoring the sense of empathy ultimately needed to rehumanize the Other: "As the members of each group describe the pain and suffering of their group at the hands of the other, they can begin to open up to the pain of the other. They can grieve for themselves, for the other, and assume responsibility for their share in the historical antagonism and violence" (1999, p. 255). At the same time, however, it has to be emphasized that this effort is built on the notion that the other individual is so complex that it is impossible or undesirable to fully identify with him or her (Halpern, 2001). For this purpose, Salomon and Nevo (2001) suggest that issues related to whether empathy is equally applicable to the victim and the perpetrator have not been adequately addressed in the peace education literature.

To sum up, reconciliation and empathy can be understood as both psychosocial and pedagogical interventions (Lockhat & Niekerk, 2000) that aim to heal the effects of traumatic events but that also involve a struggle toward relationality, often out of contexts of fear, hatred, and violence (Frow, 1998). Relationality, as I have outlined, involves opening up reflexive spaces for listening to the Other and for reimagining community in terms of inclusion rather than exclusion (Burbules, 2000; Suga, 2006; Venn, 2005). Reconciliation and empathy confront what is unspeakable or irreparable and give rise to complex difficulties of narrating and hearing trauma narratives (Humphrey, 2000).

Strengthening Empathy and Reconciliation in Education Programs

In the preceding section, I have suggested that resonating with another person emotionally is an important step toward his or her rehumanization and the subversion of polarized trauma narratives. The challenge that makes educators' work hard is the extent to which it is possible to nurture empathy. On the one hand, I share the view that while empathy and reconciliation do offer alternative narratives for the promotion of mutual understanding and peaceful coexistence, education alone cannot do much for reconciliation unless there are significant structural changes. On the other hand, the question remains: How can educators make a difference in shifting the legacies of chosen trauma? How can educational programs help the young generations transcend the politics of hatred and engage in reconciling relationships with the enemy? Having raised these questions, I now want to consider some ways at the micro level that may strengthen empathy and reconciliation in peace education curricula and pedagogies. Despite the immense challenges, schools can do much to socialize the young generation to live in peace with adversary groups (Burns & Aspeslagh, 1996; Cole, 2007; Salomon & Nevo, 2002; Wintersteiner, Spajić-Vrkaš & Teutsch, 2003).

First, reconciliation needs to take place at the emotional level (McKnight, 2004) and begins by seeing the world from the other's perspective. Educators in divided societies must do whatever is possible to nurture this emphasis through corresponding peace education programs that involve students in active learning, trauma healing, meaningful social interaction between communal groups, intercommunity cooperation, conflict resolution, and civic society education (Danesh, 2006; Gallagher, 2004; Harris, 2004; Harris & Synott, 2002; Salomon & Nevo, 2002). It should be acknowledged that each other's narratives of trauma are real events, but they are also constructed and learned and as such they may be deconstructed and transformed (McKnight, 2004). Often these narratives, especially in conflict-ridden areas, leave little or no space of empathy for the Other's losses. Thus, it is important to initiate reconciliation practices by validating each other's stories of loss and trauma. Educators must help students see the events through the enemies' eyes, if there is hope to tolerate disagreement.

In addition, Bekerman and Maoz (2005) suggest that goals such as peace and coexistence education may be better achieved if the emphasis on separate identity and culture is somewhat relaxed. According to them, strengthening coexistence might not be achieved if alternative options to the ones dictated in the past are not pursued. Educators and students should learn to be open to the possibility of transformation and the exploration of multiple ways of connecting with others. Such connections will constitute a *third space*—a space that opposes nationalist sentiments and polarized trauma narratives and opens possibilities for reimagining the sense of community and identity. An important way that pushes such connections is to avoid becoming enclosed in past identities that have been historically associated with nationalism and struggle to invent a democratic citizenship that critically reconsiders past feelings of belonging (Svašek, 2005).

Educators have a difficult task because they have to be able to help students appreciate the unknowableness of the Other. Any attempt to categorize the Other (e.g., as *barbarian,* or *evil*) in order to diminish him/her is to *totalize* him or her (Levinas, 1969). Thus, *unknowing* the Other is primary to empathy and reconciliation as well as to the process of rehumanizing the Other (Zembylas, 2007). In this manner, students may begin to approach the Other differently and invoke the value of giving up their positions as righteous. Once educators and students listen to and acknowledge the unknowableness of the Other, they can begin working toward developing a shared history and recover some of their sense of connectedness to each other, their communities, and the world. As they become less defensive, they will be more receptive to other educational programs such as human rights education, media literacy, history education, and education on stereotyping and prejudice (Cole, 2007; Danesh, 2006; Stovel, 2000).

Educators should also try to call attention to the rhetoric of binary opposites and the hegemonic ways of thinking and feeling about chosen traumas and chosen glories, and they should interrogate the manner in which they appear to be natural or common sense (Petonito, 2000). In other words, peace education curricula and pedagogies should ask students and teachers to radically reevaluate the affective economies in which they live. This process engages them in a politics that subverts emotional impasses and generates relationality and common understanding. In schools, there are numerous instances of compassion and mutual understanding that often go unnoticed on an everyday basis—these include friendships and relationships across national, racial, religious, or other divides. Teachers and students can identify and nurture such instances and consider how these moments can become inspiring for action. What is important to note is the ethical and political impact of peace education practices that involve the explicit mobilization of affect. These practices are involved in the creative production of affections (e.g., compassion, tolerance, kindness) and provoke teachers and students to participate in community struggles that refuse to accept exploitation of trauma narratives.

Finally, a powerful pedagogical strategy that may be used to strengthen empathy and reconciliation in peace education curricula and pedagogies is *critical emotional literacy* (Boler & Zembylas, 2003; Zembylas & Boler, 2002). Critical emotional literacy is a form of critical emotional praxis that focuses on questioning cherished beliefs and presuppositions. This strategy requires that individuals step outside of their comfort zones and recognize what and how one has been taught to see (or not to see). Developing critical emotional literacy means analyzing and critiquing the ways in which the politics of trauma encourage certain emotions and ideologies and prohibit others. The difficulty of developing critical emotional literacy is finding out how emotions are manipulated by the politics of trauma. This idea creates two primary challenges for educators: First, one has to deal with the polarization of trauma narratives and what they teach; second, one has to learn how to problematize the ways in which the politics of trauma serve as a form of cultural/emotional pedagogy and/or hegemony.

The affective aspect is crucial to understanding self/other relationships in peace/coexistence education programs and pedagogies, because it is emotional attachments that construct our images and identities of both self and other. These emotional attachments and landscapes need to be scrutinized so that we understand how and why we *see* some things and not others. When examining the history of traumas and conflicts, peace/coexistence education programs and pedagogies must explore the field of emotion, because its examination enriches our modulations of otherness. Critical emotional literacy situates itself within a critical narrative that emphasizes how educators and learners learn to trace their subjectivities as constantly shifting. Critical emotional literacy also entails creating spaces for epistemological and emotional problematizations of individual and collective emotions, collective historical memories, and one's sense of self and encourages an atmosphere of critical respect for difference and inquiry into the nature and implications of the politics of trauma. Finally, critical emotional literacy as a critical tool is one way of affirming the historical/political agency of the learner, because it enables him/her to formulate emotion as important knowledge in relation to culture, education, and resistance.

The ideas discussed in this chapter explicate the educational challenges to empathy and reconciliation inherent in the ethical and epistemic interconnectedness of individuals and communities. Educational programs and pedagogies must be designed, therefore, to help children become aware, both at an emotional and an intellectual level, of the shared meanings, visions, and ethical interdependence that can constrain as well as promote understanding and communal interaction. These shared meanings and visions are embodied in gestures, languages, beliefs, foods, narratives, and rituals (Cohen, 1997). Pedagogies designed to help children make choices about how they wish to relate to these shared meanings must help them overcome emotional resistances to change, and, therefore, must engage them bodily, emotionally, intellectually, and spiritually (McKnight, 2004). It is in this manner that the ideas expressed here can contribute to what Danesh (2006) calls an integrative theory of peace education, that is, an understanding of peace education that creates simultaneously a psychological, social, political, ethical, and spiritual space for teachers and students.

In this regard, peace education practices can actively facilitate the efforts for peace, coexistence, and reconciliation by helping to dismantle the system of entrenched myths and antagonistic trauma narratives that perpetuate divisions between communal groups. In particular, peace education programs and pedagogies that challenge hostile trauma narratives may offer two important things. First, they can provide a space where educators and students may question taken-for-granted assumptions and the politics of hegemonic trauma narratives. Second, these programs and pedagogies can also offer opportunities for traumatized students to work through feelings of trauma and rehumanize the Other; thus, teachers and students can begin to confront the ideological and political aspects of chosen traumas.

In educational sites following violent conflicts where suffering has been experienced by all communities, educators may choose to use the experience of one's own suffering to enhance his/her understanding of the suffering of the Other. This is not an easy task, especially because our enemies are often implicated in our suffering (as we are in theirs). Suffering, in itself, does not necessarily lead to reconciliation or empathy; however, these are attitudes that can be nourished (Cohen, 1997). Through critical educational practices, our own experiences of suffering can be used to subvert moralistic and sentimental reactions and enhance our capacity to form wise and compassionate responses to the suffering of others; essentially, such practices can help us take a critical stance toward *our* constructions of reality and the fixation on *our* own traumas while ignoring the traumas of others.

Chapter 3

The Affective Politics of Hatred in a Conflict-Ridden Area

Hatred has always been a powerful emotion in the history of humanity. Recently, however, an increasing number of scholars from different disciplines suggest that in many parts of the world tumultuous political, religious, and economic developments have generated strong emotions of hatred (Ahmed, 2001; Blee, 2004; Moore & Aweiss, 2002; Yanay, 2002a, 2002b). Thus, hatred is now widely used "in lieu of words such as violence and aggression" (Yanay, 2002b, p. 53); this shift from aggression to hatred in recent times, writes Yanay, suggests that "people search for inner motives to explain acts they find hard to conceive and forgive" (ibid.).

Previous research and analysis on hatred has largely focused on the role of hate feelings through a range of psychological and psychoanalytic discourses that consider such feelings primarily as individual and private (Yanay, 2002b; see also chapter 2). However, emotions are increasingly being recognized as part of everyday social, cultural, and political life (Lupton, 1998; Williams, 2001). The notion of politics of emotions (Abu-Lughod & Lutz, 1990) that has already been discussed in the previous two chapters represents the idea that emotions are not only individual but also political. Taking off from the work of the late anthropologist Michelle Rosaldo, and building on the ideas of Michel Foucault, Abu-Lughod and Lutz assert that what people feel physiologically as emotions is generated in social, political, and cultural encounters. Power relations, cultural principles, and rules of behavior, together with intentions others seek to implement, structure how people experience themselves as emotional and social beings and largely influence how individuals understand and talk about emotions.

In particular, recent theorizing in the sociology of emotions and sociological studies of hatred and ethnoviolence have begun to view hatred as a social product that is constituted in a political space in which individuals interact with implications in larger political and cultural struggles (Pertersen, 2002; Scheff, 1994). According to this analysis, hatred is circulated within affective economies (Ahmed, 2004) in which emotions do not reside in individuals but wherein they circulate in relationships of difference. It is also argued that within such a social and political space hatred is *ambivalent* (Yanay, 2002a) in the sense that it is full of contradiction and uncertainty and is not (always) absolute. The argument of hatred as ambivalence, asserts Yanay, enables us to locate the study of hate feelings in both individual human needs and motivations, as well as in social and political struggles.

In this chapter, I argue that using the notion of ambivalence in understanding a situation of ethnic hatred can be helpful for educators who are struggling to find pedagogical openings that deal with the powerful affective investments of hatred. Hatred has rarely been examined independently in educational research; in fact, many studies seldom make any differentiation between hatred and aggression or prejudice. However, it is extremely valuable to analyze the complexity of hatred as an emotional experience and consider its implications in education. Thus, I suggest that although hate feelings are difficult to change, unraveling the ambivalence in the affective politics of hatred creates possibilities for enriching educators' and students' perspectives on mutual understanding and solidarity with others. Drawing on insights from an action research project that I conducted in the context of Cyprus, I will show how the notion of hatred as ambivalence constitutes an important step of initiating a subversive analysis. My overall argument rests on the idea that educators need to develop pedagogies that account for the intersections of affective connections, power relations, and community-building (Zembylas, 2007). The importance of this idea rests on the notion that educators and students can extend the ethical and political entanglements of affectivity to create openings in the classroom space in which otherness and difference can be felt and articulated.

THE MEANINGS OF HATRED

In research and theoretical accounts, hatred has generally been understood in two major ways: as an individual psychological phenomenon and as a social product (Blee, 2004). Studies based on the former way stem mainly from the theoretical frameworks of psychoanalysis (e.g., see Baures, 1996; Gabbard & Winer, 1994; Klein, 1962), while studies that are based on the latter way theorize hatred as an emotional experience constituted through power and discourse—an emphasis that has been generated from recent studies on racism, nationalism, sexism, and colonialism (e.g., see Pertersen, 2002; Scheff, 1992, 1994; Wetherell & Potter, 1992).

Hate as an individual phenomenon is understood as an aggregate of beliefs, values, and affects within an individual who is negatively positioned with respect to those who differ from him/her. According to many psychoanalytic theorists, hate develops in early infancy as a means of resisting threats and protecting one's self from pain and danger; it is then reproduced in adulthood as unhappy, frustrated, and dysfunctional relationships (Klein, 1962). In other words, hatred signifies a defensive mechanism of emotion toward self-preservation and protection, especially when social, economic, political, and other psychic conditions are perceived as threatening (Blee, 2004). Thus, in the psychoanalytic understanding of hatred, hate feelings toward other ethnic, racial, and cultural groups are perceived as fundamental emotional features of the prejudiced personality (Allport, 1954). Prejudice and discrimination are assumed to be universal psychological blocks rooted in basic human desires for power, status, and social connection such that people tend not to value those who are dissimilar to themselves (Aronson, 1992). In this sense, hatred sets clear boundaries that distinguish *us* from *them*.

Although considering hate as rooted in individual attributes has been useful for analyzing why certain individuals become violent against others, such an individualistic focus has limitations in analyzing a social phenomenon, because there is the danger of reducing a social phenomenon to "a psychological, ego-bound" event (Yanay, 1996, p. 23). This approach "tends to strip individual action away from a social context, thereby assuming as universal certain sentiments, emotions, and ways of thinking that might, in fact, be variable across time and place" (Blee, 2004, p. 98). Consequently, it has been argued that the psychoanalytic perspective constitutes a weak explanation for the collective hatred of a group against another group, especially in cases of racial, ethnic, or religious hatred (Harré, 1986).

In their studies on ethnic conflict and ethno-nationalism, Scheff (1992, 1994; Scheff & Retzinger, 1991) and Retzinger (1991) argue that hatred for others could be a defense of a group against hiding feelings of shame and alienation under the cover of ethnic pride. Thus, the most powerful way of forming an *us* and *them* mentality is to idealize one's own group and demonize the Other—an idea that alludes to the political aspects of hatred. Idealization and demonization are accomplished through *myth-making*—accounts that justify the negative evaluation of other groups and glorify one's own nation (Aho, 1994). In this sense, there is no recognition of heterogeneity among the Other (Blee, 2004); for example, all Turks or Greeks, all Muslims or Christians, all gays or lesbians, all blacks or whites, are rendered the same through feelings of hatred. In essence, each group *delegitimizes* (Bar-Tal, 2003) the actions and beliefs of the other and regards all members of the enemy group as "evil, malevolent, immoral and inhuman" (ibid., p. 83).

The implication of a relational and political understanding of hatred is that it shifts attention from the individual to the practices with which hate is socially, historically, culturally, and politically constituted (Aho, 1994; Blee, 2002). In other words, it allows for a theorization of the affective as one among several economies in society such as capital, money, identities, meaning, representations, and desires (Grossberg, 1988). Economy here is used in the sense of "circulation and distribution of energy" (Grossberg, 1988, p. 283). This idea implies that affect is one of the ways through which power is constituted, circulated, and performed. Thus, an economy of affect indicates how people give and withhold emotional resources, form social relationships and differences, negotiate meanings, and constitute their subjectivities (Harding & Pribram, 2004).

The notion of affective economies (see chapter 2) is particularly useful in theorizing an economic understanding of hate, because it emphasizes that hate does not reside within an individual but is circulated (Ahmed, 2004). In addition, this notion shows how hatred draws other bodies together making them members of a group united by their hatred to other groups. This economy of hate, asserts Ahmed, works to differentiate some bodies from other bodies. In particular, there are two important advantages in considering the notion of affective economies to explore practices of hatred. First, an economic understanding of hatred allows us to see that affective connections of hatred involve movements, intensities, and energies. Differences among bodies are materialized as relationships and affective encounters; they are not inherent to a particular body. Thus, hatred is an effect of the affective encounters within a community, and it is through moving toward (and away) from others, objects, or signs that bodies create hate feelings and become aligned with or against other bodies. Second, a theory of hatred as economy involves relationships of difference and thus is a political theory that does not conceal the traces of differentiating some bodies from others. In this theory, individuals do not inhabit the skin of the community, as Ahmed (2004) says, but rather the skin of the community is an effect of the affective economies. Thus, hatred aligns not only the *I* with the *we* but also the *you* with the *them* (Ahmed, 2000).

Ahmed (2000, 2001, 2004) follows Matsuda's (1993) work and emphasizes the importance of the affects of hate on the bodies of the victims. Therefore, in economies of hate, hatred *does things,* as she explains, that is, it produces affects on the bodies of those who are its objects. As she writes: "Hate is not simply a means by which the identity of the subject and community is established (alignment); hate also works to unmake the world of the other through pain" (2001, p. 360). The bodies of the victims, then, are *testimonies* of hatred, because the bodies of others are transformed into *the hated* (Matsuda, 1993).

The above ideas urge us to consider how hatred organizes the social and bodily space creating a powerful emotional border. The alignment of bodies with some and against others takes place through the affecting of movement; bodies move away from some bodies or close to others who are recognized as hated or loved, respectively. The bodies of others are therefore transformed into the hated through a discourse that assumes they cause pain and injury to us. Hate is distributed in such discursive practices that come to signify the danger of impurity from mixing with *them;* they threaten to violate our purity. Thus, it is the emotional response of hate that works to constitute who the victim and the victimizer are.

Hatred as Ambivalence

The separation of others into bodies that can be loved and hated also points to the ambivalence of emotions, that is, the notion that love and hate are not attributes of bodies or preexisting emotions (Ahmed, 2004) but practices that recognize similarity and difference in discourse and culture. In hating others, one also loves those who are threatened by those others; thus, the attachment to some bodies is divided as negative and positive (hate and love) through imaging others who are unlike/like us. As Ahmed (2000) writes: "If the demand for love is the demand for presence . . . then hate and love are intimately tied together between desire and loss, presence and absence. To some extent, hate is an affect/effect of the impossibility of love. . . . Hate, then, is tied with the lack that is concealed by presence and revealed in the demand for presence" (p. 352). This rather psychoanalytic interpretation emphasizes the ambivalence of emotion discourse in that the language of hatred is not stable. Turning away from others simultaneously moves us toward those who are like us; in other words, some forms of hate are expressions of love.

Weigert and Franks (1989) write that ambivalence essentially means "I cannot decide what to do; I do not know what to think; I have mixed feelings about this" (p. 206). As a construct, ambivalence originated within the psychoanalytic movement. Freud used it to refer to the experience of simultaneous contradictory feelings, especially in relations of intimacy, of narcissistic tendencies, and of taboos (Merton, 1976). However, Merton moved ambivalence from its psychoanalytic use and introduced its sociological dimension of contradictory pressures in social roles. Ambivalence, then, can be used in two different, but related and complementary meanings: the psychoanalytic notion of *contradiction* and the sociological notion of *uncertainty*. On the one hand, ambivalence refers primarily to the coexistence of two opposite feelings; on the other hand, this acknowledgment of difference goes beyond the binary logic of an either/or schema. The latter interpretation allows for the simultaneous presence of opposed emotions and does not always demarcate the meaning of *us* and *others* in absolute terms. In other words, ambivalence is considered a situational or contextual concept, primarily because collective relationships are frequently marked by changing nuances and interpretations (Weigert, 1991).

More recently, ambivalence has come to refer to uncertainty and continuous movement because of the increasing complexity and indeterminacy of the (postmodern) world (Bauman, 1991). In particular, poststructuralist writings have revived interest in the notion of ambivalence. This interest is based on the idea that power relations are ambivalent and that a hegemonic event (e.g., nationalism) not only excludes and creates inferiority and conformity, but also mobilizes *subversive* resistances (Bhabha, 1994; Butler, 1997b). For example, based on how hate feelings as emotional practices can preserve communication (even if negative), it may be argued that hatred as a form of power is an ambivalent mode of knowledge (Bhabha, 1994; Butler, 1997b). It is precisely because of this potential that Butler (1997b) theorizes ambivalence as a site of subversion. She writes that the ambivalence between presences and absences makes the naturalness of hegemonic norms forever unstable. Similarly, Bhabha (1994) asserts that power relations are constituted in ambivalent feelings such as love and hate and are never simply polar opposites but hybrid locations in discourse and culture. Thus, by pointing to ambivalence, there is an attack on attempts to present emotional or political practices as naturalized and therefore as stable.

The notion of ambivalence as hybridity challenges both the psychoanalytic (Freud) and the sociological (Merton) interpretations of ambivalence based on the idea that there are various degrees of oppositional elements and multiple meanings as a result of the instability of discourse (Derrida, 1978). In other words, there are multiple meanings in discourses of hatred; the language of hatred is not unitary (Yanay, 2002a). The instability of discourse acknowledges the importance of the performative aspects of affect (Zembylas, 2005), because ambivalence is conceived as constituted within emotional practices. A turn away from the psychoanalytic and sociological notions of ambivalence, the notion of ambivalence as hybridity opens up possibilities to use emotional practices as means through which solidarity may be constituted (Bauman, 1992) and social change may be promoted (Bhabha, 1994).

AMBIVALENT EXPRESSIONS OF ETHNIC HATRED: THE CASE OF CYPRUS

To show how hatred as ambivalence may be used constructively in the context of education, I will offer some insights gathered during an action research project over a three-year period (from October 2003 to May 2006) in a higher education institution in south Cyprus.[1] The purposes of this project were: to understand the emotional dynamics of hatred in my students' (Greek-Cypriot preservice teachers) narratives about the political situation in Cyprus[2]; and, to explore new possibilities of practice and thought, along with respective resources and strategies to deal constructively with feelings of hatred. In this chapter, I focus my discussion on identifying the different modes of hate feelings among my students, analyzing how those expressions of hatred constitute ambivalence and exploring the openings that are created for educators who want to promote empathy and reconciliation.

Cyprus, as a divided country, provides an interesting case study of studying the affective economies of hatred. There is considerable evidence from various studies that school textbooks and national rituals, symbols and celebrations systematically create dehumanized images of the Other within both the Greek-Cypriot and the Turkish-Cypriot communities and inspire hatred for the enemy (AKTI, 2004; Bryant 1998, 2001, 2004; Hadjipavlou-Trigeorgis, 1998; Kizilyürek 1999, 2001; Papadakis, 1995; POST Research Institute, 2004, 2007; Spyrou 2001a, 2001b, 2002, 2006). More particularly, there is much ethnographic evidence indicating how individuals as well as organized groups from both communities systematically attempt to nationalize suffering and highlight the need to remember what the enemy has committed in the past (Bryant, 2004; Loizos, 1998; Papadakis, 1998; Sant Cassia, 2006). Further, each community constructs national monuments, which not only embody the losses and hurts suffered but also represent the accumulated grief and hatred between Greeks and Turks over the centuries.

Thus, it is argued that the intensification of national hatred in Cyprus shapes a sense that there is a commonality in being Greek Cypriot or Turkish Cypriot with Greeks and Turks, respectively (Attalides, 1979; Kizilyürek, 1993; Mavratsas, 1996, 1999). In fact, school education promotes the use of more inclusive categories such as Greeks or Turks, at the expense of more synthetic or hybrid ones such as Greek Cypriots and Turkish Cypriots (Spyrou, 2006; Theodossopoulos, 2006). In viewing hatred as relational and political, each group constructs its ethnic identity through learning to hate the Other. Not surprisingly, then, memories of life before the conflicts between the two communities—that is, memories of friendly interethnic neighborhood relations—become part of a hidden discourse, because they are at odds with the official nationalist discourse (Loizos, 1998). The nationalist discourse of the suffering nation becomes a powerful image of hatred by evoking emotional memories of suffering from the past (cf. Duijzings, 1999; Ignatieff, 1995). This victim ideology is a powerful defensive mechanism that justifies aggression and hatred against the Other, establishes the illusion of coherence and order, and masks the anxiety of seeing any similarities with the Other.

Overall, although the long-standing impact of ethnic nationalism on collective identity and memory has generally been analyzed in Cyprus, it is also interesting to examine the affective economies of hatred and how they work in the context of prospective teachers' beliefs. This is a particularly important issue not only because these individuals will become educators but also because this age group of individuals has grown up in the south after the events of 1974 and thus the Turkish-Cypriot population is virtually unknown to them. This is not to say that the insights offered here constitute an in-depth analysis of the emotions, beliefs, and perceptions held by Greek-Cypriot prospective teachers on the situation in Cyprus. Far from that, this analysis focuses on expressions of ethnic hate speech illustrating that not all hate speech is the same but that there are multiple and ambivalent expressions of hatred (Boler, 2004; Yanay, 1995, 1996, 2002a).

In order to show the different expressions of hatred, I will analyze some exemplary quotes that were compiled from interviews with four students, and then I will discuss how hatred is manifested in these students. The interviews—formal and informal—took place at the beginning stages of the course that I taught. These quotes were representative of the four major positions that were repeatedly identified in the three-year action research project. These positions range from those who acknowledge hatred and feel it is legitimate to have these feelings because of what Greek Cypriots suffered in the past, to those who do not acknowledge that they feel hatred, or to those who acknowledge they have clearly ambivalent feelings.

A: I don't really care how this sounds but I don't want to live with the Turks! I hate them! We suffered so much because of what they did to us in 1974 and we still suffer from that. I don't want them here. Period! [. . .] I think it's better if we live separately. I have such strong feelings about this that when I see Turks here [after the opening of the Green Line] I feel anger and disgust and I want to kill them! [. . .] Yes, I feel afraid because I think they want to conquer Cyprus and finish what they started back in 1974. Their intention is to make Cyprus a Turkish territory. (male, sophomore student)

B: I don't think I am a nationalist. OK, the Turks are our enemies; they are occupying our land, so I think it's only natural for us to feel this hatred against them. I don't see anything wrong with that. I just wish they left and let us live peacefully. We want peace; they want war . . . Why don't they just let us live peacefully and go back where they came from? It's not fair, you know? This is our land! Cyprus has always been Greek. [. . .] I have no doubt the Turks hate us and would kill us all, if they had the chance. Barbarism is in their blood since they first appeared on earth. They have always been like that. I mean this is historically documented. Think about what happened in Constantinople, in Asia Minor. . . . (female, junior student)

C: I am aware that we grew up hating the Turks. The schools, the church, the society certainly contributed in creating this perception. But I think it's impossible to overcome this feeling of bitterness and rage. That's just how I feel about it. Perhaps my feelings will change in the future. . . . I know I shouldn't make generalizations and say that all the Turks are bad people. But every time I make an effort to see things differently, all these images come to my mind: our occupied villages; pictures of our beautiful churches and historical monuments that the Turks took from us; images of killings and destructions that we suffered during the Turkish invasion; our refugees. . . . With all these images haunting us, how can we ever live with them? I just hope that one day I'll be able to overcome these feelings. But right now, that's just how I feel. (male, junior student)

D: No, I don't hate the Turkish Cypriots. I don't have anything against them. I despise the Turks who invaded Cyprus and I despise the Greek Cypriot "super-patriots" who caused the military coup and gave Turkey the opportunity to invade Cyprus. [. . .] Both the Greek Cypriots and the Turkish Cypriots suffered a lot. We caused much suffering to them during the period 1963–1974 but nobody talks about this on our side. So I see their perspective too. They've been refugees, they've had missing persons too, and they've lost loved ones. [. . .] After the opening of the Green Line in 2003, I realize more and more how the Turkish Cypriots look like us! I mean it's hard to tell if someone is a Greek Cypriot or a Turkish Cypriot! Can you imagine my surprise when I think someone is a Greek Cypriot and then I hear them speak in Turkish? [. . .] They are human beings just like us. They are not the monsters we have been taught in our schools, by our politicians for such a long time. . . . [. . .] Yes, despite what happened in the past, I think we can live together peacefully respecting each other, because Cyprus belongs to all the Cypriots. I hope that this will happen soon. (female, sophomore student)

Clearly, the positions of the first two students (A, B) differ from those of the last two (C, D). The first two positions are largely intolerant of the Other; it is also interesting that no distinction is made between the Turks and the Turkish Cypriots, assuming that they all belong to the same category. That is, for these two positions, the Other is stereotyped—a member of a group that is perceived as threatening. Interestingly, these same positions are reflected in the findings of Spyrou's research (2006) in which the beliefs of Greek-Cypriot elementary school children about the Other are investigated. These positions represent a nationalist, ethnocentric ideology that is embedded in a discourse of moral superiority and self-righteousness. The first position acknowledges that it is nationalistic and adopts a totalizing discourse to legitimate killing, whereas the second position is completely unreflective; all the Turks are "the enemy" and consequently they are hated for that (see also Bryant, 1998, 2001; Hadjipavlou-Trigeorgis, 1998; Loizos, 1998; Spyrou, 2001a, 2001b, 2002).

Furthermore, the first two positions provide a clear sense of the intensity of the emotional attachments to a community; hatred does not reside in the individuals expressing these positions, but it seems to slide between individuals and aligns them with the larger community ("we") that is brought together by a common threat ("them"). What separates these two individuals from others also connects them to all the Greek Cypriots who feel threatened by the Turks. The narrative of hatred functions as entitlement ("This is *our* land") as well as displacement ("they want war"; "Barbarism is in their blood"). There is an alignment of those who belong to the victimized community with good and justice ("We suffered so much"; "We want peace"); the others are hated fiercely as being evil and unjust ("They have always been like that"). It is the emotional response of hate that works to bind the individual and the imagined community together (Ahmed, 2004); in other words, hate works to constitute the community together ("I think it's only natural for us to feel this hatred against them").

Such expressions of hatred are clearly internalized by the first two individuals and signify the fear of being harmed by the Turks ("the Turks hate us and would kill us all"). This fear is based on previous events—not only recent ones but also ones that go way back in history ("Think about what happened in Constantinople, in Asia Minor"). The feeling of "disgust" as expressed in the first position indicates the effect of hatred on the body ("when I see Turks I feel anger and disgust and I want to kill them"). This statement is an explicit expression of the "rippling effect of emotions" (Ahmed, 2000, p. 347) and its visceral impact on the body (Matsuda, 1993). Finally, it is interesting to point out that even in the statements "I hate the Turks" or "When I see the Turks here I feel anger and disgust" in which the Other is rejected outright, there is ambivalence but it is completely unacknowledged. This ambivalence is based on the idea that the presence of the Other is inevitable (Yanay, 2002a).

On the contrary, the last two positions (C, D) illustrate more explicitly the ambivalence of emotion, by acknowledging that there are both positive and negative attributes regarding the Other. Both of these positions discuss feelings of hatred as well as feelings of understanding and empathy ("Perhaps my feelings will change in the future"; "I see their perspective too"). The fourth position seems more open than the third one; the third position acknowledges ambivalence and is hesitant and tormented ("I just hope that one day I'll be able to overcome these feelings"), while the fourth position is more courageous and empathetic ("both the Greek Cypriots and the Turkish Cypriots suffered a lot"). Both of these individuals discuss their ambivalence and recognize that they cannot make universal generalizations and categorizations. The third position is more emotionally uncertain than the last one, shifting from feelings of hurt to the negation of absoluteness, whereas the last one focuses on argumentation and understanding (see also Yanay, 2002a).

The discourse employed by these two individuals is open to the politics of difference and shows respect for the Other ("I know I shouldn't make generalizations and say that all the Turks are bad people"; "They are human beings just like us"). Both individuals problematize their feelings of hatred rather than denying this possibility. The ambivalence of hatred is bound up with the positing of an affective relationality between the "I" and the "we" (Ahmed, 2004). But the meaning of "we" is neither fixed nor univocal; in fact, the fourth individual becomes emotionally attached to new meanings of "we" as she acknowledges the politics of difference. Thus, understanding the Other becomes an affect and effect of identification and works to read the Other outside of stereotypes (Bhabha, 1994). The last two positions are more fluid and permeable compared to the first two positions that seem more monolithic and fixed. The question now is: How can this ambivalence—acknowledged or unacknowledged—be used to create solidarity and overcome hatred, especially in the context of a higher education classroom?

The Struggle to Cultivate Solidarity and Overcome Hatred in Education

My analysis of the affective economy of hatred has demonstrated that there are different workings of hate feelings, as those are reflected in the four preceding positions that were discussed. These insights provide further insights to recent findings on the ambivalent feelings of individuals and communities in conflict-ridden areas (Loizos, 2005; Yanay, 2002a). Furthermore, this analysis shows how the meaning of hatred is historical and political rather than simply individual and psychological. The notion of affective economies links the individual with the social and political formation through practices and energies of affect (Harding & Pribram, 2004). Thus, the affective economy of hatred becomes an important force and intensity that motivates power relations among individuals and communities. Theorizing issues of hatred from this perspective illustrates that the imaginary emotional boundaries constructed within communities are not absolutely fixed but rather ambivalent.

Consequently, the task of progressive pedagogies would be to identify practices, strategies, and spaces in which solidarity might be possible. In particular, coming to understand and define the ambivalence of hatred through dialogue and relationality (see chapter 2) is one way educators and students can produce different interpretations and understandings of how they feel about others. It is important for dialogue to emphasize the ambivalence of emotions, by showing that positive and negative are not attributes of emotions or of individual bodies, but they represent provisional readings and judgments that have profound effects at a political level (Ahmed, 2004). In this respect, educators and students can develop analyses to interrogate how affective economies of hatred establish, assert, and reinforce power differentials and hierarchies and understand how individuals draw on the resources of ethnic stereotypes to reproduce nationalist beliefs and practices. Therefore, problematizing the emotional practices of hatred, looking at how hatred circulates in discursive practices, and considering what it *does* to people are crucial for exploring pedagogical ways that could produce alternative affective connections and subvert hatred.

In particular, engaging in *dialogues of negotiation* (Burrow, 2005) in the classroom builds on the acknowledgment of ambivalence and permits similarities to be reached through acknowledging difference. Rorty (1989) argues that in order to feel solidarity with people we consider different from us, we need to find ways to notice our similarities with them and negotiate new interpretations of *we*. For example, the realization that not all the others are evil people or that they look like *us*, as pointed out by the last two positions, may be a good point of departure for a collaborative effort of interpretation and reinterpretation that emphasizes similarities without ignoring differences. The process of *relational empathy* (Broome, 1991, 1993) can also be another useful step toward the development of shared meanings created through interpersonal encounters—especially when there are opportunities to come face-to-face with the Other (Levinas, 1987a) such as in the case of Cyprus after the opening of the Green Line in 2003. The focus of relational empathy is not only on understanding the Other but also on developing shared meaning with the Other as the basis for relationship growth.

My argument in this chapter has been that hate feelings are political manifestations of various kinds not only of resistances but also of possible transformations. The language of ambivalence certainly provides an alternative discourse that "invites and invents solidarity with the needs and pain of others" (Yanay, 2002a, p. 85). Even in cases in which the workings of hatred is expressed through nationalistic and exclusive discourse, educators and students can negotiate and expose their ambivalence to provide resources for expressing, interpreting, and reflecting on their emotions. As Yanay (2002a) points out:

> Ambivalent discourse represents contradictory possibilities of speech. On the one hand, anxiety, the outcome of ambivalence, produces denial, disturbed seeing and disavowed knowledge. On the other hand, when ambivalence is acknowledged, contingent verbalization, through cognate awareness and via a process of mirroring, limits the scope and valence of hatred. (p. 82)

Unraveling the political aspects of hatred—both as analytic tools and as points of departure for cultivating individual and collective political consciousness and critical hope—creates possibilities for enriching educators' and students' perspectives about the dynamics of relationality in learning about/with the Other.

In conclusion, the pedagogical activities that inspire political consciousness and criticality need to include not only interrogation of the possibilities and limitations of the notion of ambivalence but also provision of a dialogical structure within which the resources are provided for expressing and interpreting old and new affective relations. Teachers and students have an obligation to be open to the possibility of transformation and the exploration of multiple ways of connecting with others (Ropers-Huilman, 1999). Furthermore, acting to promote the practice of such connections—such as intimacy, kindness, and compassion—helps teachers and students create movements of difference and hope that can act to propel mutual understanding, empathy, and reconciliation. Undoubtedly, understanding how hatred works in different ways demands an acknowledgment of the affective ways in which people are involved in their realities. Individuals and groups have powerful affective investments in their identities that cannot be ignored but need to be respected (Albrecht-Crane, 2005) in any pedagogical effort that encourages empathy and reconciliation. In the context of education, this view implies an entirely new ethic that highlights the importance of cultivating affective relationality such as mutual respect, solidarity, and compassion. Ethics as sensibility demands an openness to the affective potentiality of encounters with others—encounters from which new ways of feeling, thinking, and being might emerge.

Chapter 4

Understanding Fear

Discourses of fear and anxiety circulating in the United States after September 11, 2001 and responses to immigration and asylum seekers with a similar discourse type in the European Union draw significant boundaries among *us* and *them*—that is, the *friends* from the *enemies*, and the *citizens* from the *non-citizens*. Work in the cultural politics of emotion teaches us that we cannot think of the issue of national identity or community—of what it means to be a member of a nation or a group—without reference to the affective origins of individuals and their communities (Ahmed, 2000, 2004; Kristeva, 1993; Lupton, 1998). This work suggests that emotions play a fundamental role in the constitution of the relationship between individuals and communities.

In particular, the *affective politics of fear*, as Ahmed (2004) calls the intersection of affect, fear, and politics, *does* something extraordinary; it establishes a distance between bodies that are read as *similar* and those that are considered to be *different*. Fear involves relationships of similarity and difference, which are crucial to establishing how *our* bodies are differentiated from *theirs*—for example, the body-at-home (the host) who receives others, the strangers (Ahmed, 2005). The politics of fear acknowledges the important role of power relations in the process of establishing others as *fearsome*. Who gets constructed as the host who receives the stranger is an effect of relations of power. Thus, affects—for example, anxiety and fear for the stranger—work by establishing boundaries; the other is fearsome because he/she is constructed as a danger to "our" very existence.

Alignments between individual and national bodies have important implications for the politics of trauma in education, because the *culture of fear* (Glassner, 1999) provides an influential affective orientation for the educational system of a society (Boler & Zembylas, 2003; de Rivera & Páez, 2007; Zembylas, 2007). A number of studies in recent years have pointed to significant changes in American and European fears, in ways that cry out for assessment of the impact of fear on the politics of trauma in general (Bauman, 2006; Bourke, 2006; Füredi, 2002; Robin, 2004) and on schooling more specifically (Giroux, 2003). This work indicates how American and European schools have been caught between reproducing the hegemonic discourses of fear and trauma and approaching their affective politics critically. In light of the politics of fear, educators in the United States and the European Union cannot afford to ignore questions such as: How do existing educational practices (e.g., raising the flag as a patriotic declaration of national love; forbidding the Muslim scarf in schools) contribute to perpetuating feelings of fear and anxiety about *others*, who may be labeled as suspects of terrorism or evil? How can schools engage in a critical interrogation of the forms of inclusion and exclusion that are established through hegemonic discourses of fear in the United States and the European Union?

In this chapter, I want to offer some reflections on the above questions through an exploration of the affective politics of fear in education and to discuss the implications for educational policy, theory, and practice. The aim is to highlight the ways in which fear, in its unavoidable invasion to the pedagogic field, *educates* students to create particular affective relations to one's self and to others. In this manner, I argue that narratives of fear function as influential *pedagogies of emotion* (Worsham, 2001), and as such they are particularly effective ways of placing teachers and students within a conflicting ethos of life. Analyzing the implications of the affective politics of fear in the context of education is important for educators who want to establish a different *affective community* (Gandhi, 2006) in the classroom, a community that is not paralyzed by fear and trauma but deconstructs arbitrary boundary formations and encourages an ethos of *conviviality* (Gilroy, 2004).

My analysis is done through argumentation and the evidence of some research examples from my own ethnographic work both in the United States and in Cyprus[1]; the aim is not to generalize the role of fear in the conservation of existing power structures in schools but simply to show how fear may have important implications for the politics of trauma in schooling. Based on this analysis, I draw out some lines of ethical and pedagogical potential that emerge through such an effort, and extend the field in which pedagogues might move to induce critical resistance to the affective politics of fear. My argument is that if educators are committed to inspire individual and social change—a change that would eventually interrogate the symbolic violence exerted by the affective politics of fear—then much work needs to be done at the affective level.

THE AFFECTIVE POLITICS OF FEAR

Before considering the relationship between fear and politics, two general clarifications are worth mentioning concerning fear. First, there is a useful distinction between fear and anxiety; fear has an object (i.e., the threat is identifiable) whereas anxiety does not usually have one (i.e., the threat is vague) (Rachman, 1998). Second, fear is at its most fearsome when an object approaches and thus involves an anticipation of hurt or injury (Bauman, 2006). The threat that the object of fear approaches but is nowhere to be seen—its presence qua its absence—intensifies fear. As Bauman explains:

> Fear is at its more fearsome when it is diffuse, scattered, unclear, unattached, unanchored, free floating, with no clear address or cause; when it haunts us with no visible rhyme or reason, when the menace we should be afraid of can be glimpsed everywhere but is nowhere to be seen. "Fear" is the name we give to our *uncertainty:* to our *ignorance* of the threat and of what is to be *done*—what can and what can't be—to stop it in its tracks—or to fight it back if stopping is beyond our power. (2006, p. 2; original emphasis)

Bauman conducts an interesting genealogy of fear in the premodern and modern era and argues that in premodern times "fear was everywhere" whereas at the beginning of modern times there was a perception that the fear pervading social life could be left behind. However, as he shows, fear still reigns in the twenty-first century; whether it is the fear of natural disasters, the fear of environmental catastrophe, or the fear of indiscriminate terrorist attacks, there is a prevalent culture of fear in all aspects of contemporary life. In particular, Bauman analyzes the notion of *derivative fear*, a sort of "second degree" fear that is best described "as the sentiment of being *susceptible* to danger" (2006, p. 3; original emphasis), a feeling of insecurity and vulnerability.[2]

Robin (2004) also provides an exciting analysis of fear as a political idea. He argues that historically political fear has operated in two ways: First, politicians or militants may define or describe what they feel ought to be the public's chief object of fear. Second, political fear arises out of the socioeconomic and political hierarchies that divide a society; fear uses intimidation tactics to make sure that one group dominates another. In both forms of fear, the society is mobilized to affirm its values in ways that highlight what divides a group with the "rival" group so no windows of rapprochement are left open. The idea of fear as one of the most powerful mechanisms through which the structures of society are transmitted to individual psychological functions has been established by Elias (1982). As Elias notes, "the strength, kind and structures of the fears and anxieties that smoulder or flare in the individual never depend solely on his own 'nature.'" Rather, they are "always determined, finally by the history and the actual structure of his relations to other people" (ibid., p. 327). Altheide (2002) writes along the same lines that "fear does not just happen; it is socially constructed and then manipulated by those who seek to benefit" (p. 24).

Fear, as an affective politics, then, is not an emotion that comes from individuals and is then directed toward others (Ahmed, 2004). Rather, fear works to align individual space with social space. Fear produces *fearful* subjects in relation to *fearsome* others and secures the very boundaries between us and them. Thus, fear does not reside in a particular object or sign. This lack of residence allows fear to slide across signs and between bodies. This sliding, argues Ahmed (2004), becomes stuck only temporarily, in the very attachment of a sign to a body, whereby a sign sticks to a body by constituting it as the object of fear (e.g., the Muslim scarf or the Black immigrant from a poor African country). Fear creates boundaries between "what I am" and "that which I am not," through the very affect of *turning away* from an object that threatens "that which I am."

The response of fear is dependent on particular narratives of what and who is fearsome; consequently, "the more we don't know what or who it is we fear *the more the world becomes fearsome*" (Ahmed, 2004, p. 69, original emphasis). This is an important aspect of the affective politics of fear, because the world itself becomes a space of potential danger. Fear, then, is bound up with how we inhabit the world "with" others (Ahmed, 2005). Fear works to enable some bodies to inhabit and move in public space through the restricting of other bodies to spaces that are enclosed, such as when nation-states create policies to prevent illegal immigrants or bogus asylum seekers from entering the state. It is the flow of fear that establishes these boundaries—the fear that illegal immigrants and bogus asylum seekers, for example, threaten the well-being of a state or the character of a nation. The possibility that we might not be able to tell the difference of others' bodies—that is, between a legal and an illegal immigrant, a genuine and a bogus asylum seeker, a Muslim terrorist and a Muslim victim of terrorism—converts into the possibility that *any* of those bodies may be fearsome criminal intruders.

Narratives of fear about strangers and intruders are formed through the ways in which fear circulates to produce differentiations between us and them, whereby the them is constituted as the reason or the justification of feelings of fear (Ahmed, 2004). I want to build on this argument by suggesting that it is through the circulation of hegemonic discourses of fear that individual bodies become aligned to those considered as part of the same community, and against others that are stereotyped as the enemies of this community. Importantly, this alignment on the basis of similarity/familiarity and difference/strangeness is historically constituted in the context of specific power relations. The organization of social and affective community creates norms of exclusion and inclusion; this differentiation is never over, as new *others* are always arriving.

Clearly, to say that fear is crucial to the formation of boundaries is also to suggest that what makes those boundaries also unmakes them (Ahmed, 2005; Zembylas, 2007). That is, what separates us from others also connects us to others, if the existing narratives of fear are somehow critically interrogated and emotion discourses on immigration, asylum seeking, and terrorism are carefully rearticulated. In other words, these boundaries are not fixed but change all the time. Hegemonic discourses of fear coexist with antagonistic discourses (e.g., discourses of conviviality, solidarity, empathy, and reconciliation); thus, by deconstructing hegemonic narratives of fear, one can grasp part of the complexity and the potentiality of alterative discourses about others.

GLOBAL ECONOMIES OF FEAR

A basic idea that has been established so far is the notion that emotions do not reside in subjects or objects but are produced, circulated, and capitalized on to achieve political purposes. An affective economy, to use Ahmed's (2004) term, circulates the use-value of particular affects held by hegemonic discourses to impose the legitimate mode of feeling and thinking within a community. Thus, an affective economy of fear imposes certain affective associations that stick various signs (e.g., Muslim and terrorism) together. The bodies of us and them are felt as similar or different within globality; globality itself, according to Ahmed, can be theorized as an affective economy of fear, in the sense that it becomes an affective community through the movement of some bodies but the fixing of others.

The complexity of this affective economy of fear has never been more apparent since September 11, 2001. After September 11, the fear of the other is not contained within either a single territorial entity or a given place of origin. Fear is "ambient" and "saturates the social spaces of everyday life," writes Hubbard (2003) (see also Papastergiadis, 2006). The boundlessness of this politics of fear in the "war against terror" has chilling effects in public discourses around the world: social life, especially in "target areas" such as the United States and various European countries, becomes dangerous, fearful, and filled with potential victims (Altheide, 2006). Terrorists are immediately identified as perpetrators of fear, that is, those who seek to make others victims of terrorism. Fear is now part of everyday discourse and structures the social life of individuals and communities. The structural possibility that anyone could be a terrorist is materialized especially for marginalized groups whose stereotyping is exacerbated (Altheide, 2006).

Here I want to discuss two examples that indicate the alignment of individuals with collective bodies and the exclusion of others in the global economies of fear that have been formulated in recent years. The first example is the widespread use of the American flag after September 11; this new *patriotism* that emerged sparked old debates about the meaning and value of patriotism as well as the emotions involved, including fear. The second example is relevant to the challenge of mass migration and asylum seekers in Europe and how identity politics and fear are created at the national level to justify the exclusion of strangers from civic life. As I argue, these examples and certainly many more others contribute to strengthening the politics of trauma both in society and in schooling, because they sustain existing boundaries between *us* and *them*.

For some, the widespread use of American flags in the aftermath of September 11 may represent a nationalist outbreak and a show of chauvinistic military power. For others, this represents an incredible feeling of sorrow over the loss of innocent people and a need to respond to the politics of fear by creating some form of solidarity (Zembylas & Boler, 2002). This new patriotism invoked during the aftermath of 9/11, as Megan Boler and I argued, represents not simply an understandable reaction of grief and loss but, arguably, the ethically questionable political manipulation of fear. In the name of patriotism, fear has been used by ideological forces such as mass media to support a radical legislative redefinition of civil liberties, military and foreign policies justified by careful definitions of who counts as a terrorist, and new justifications for racism.

In other words, the politics of fear serves as a conceptual linkage for power, propaganda, news, and popular culture and thus invites policing and intervention to prevent further victimization (Altheide, 2006). As Altheide writes succinctly, "A public discourse of fear invites the politics of fear" (p. 423). Building on a foundation of fear, national character is played out in scenarios of heroics and sacrifice (Denzin & Lincoln, 2003). What is crucial here is that experiences of fear become lived as patriotic declarations of love, which construct those who do not share these experiences as being on the other side, that is, as sources of "our fear" and as reasons for new forms of border policing (Ahmed, 2004).

The events of September 11 challenge educators concerned with democracy to ask: For whom and when is patriotism a productive feeling, particularly if one cares for international peace and justice? What is the role of education in engaging students to think about the historically repeated connection between "the swell of patriotism" and the justification of military aggression against others? How does the politics of fear around terrorism end up being used against youth and their problems? As Giroux (2003) points out: "The greatest challenge Americans face does not come from crazed terrorists, but in the ongoing battle to expand and deepen the principles of justice, freedom, and equality on behalf of all citizens—especially young people, who are quickly becoming an abandoned generation" (p. xx).

It is important to recognize that the figure of the international terrorist has been mobilized in close proximity to the figure of the immigrant and the asylum seeker (Ahmed, 2004). The discourse of fear now in Europe includes not only terrorism but also the conceptualization of immigration as a security concern (Buonfino, 2004). Immigration has turned into one of the greatest concerns of twenty-first-century Europe. Terrorism and immigration play well together with audiences accustomed to the politics of fear as well as with demands for social policies geared to protect those audiences from the threat posed by *strangers*.

Immigration, in particular, is cast as a terrible trend that threatens the values of freedom, security, and culture in Europe.[3] Like the prospective victims of the international terrorist, citizens of Europe are made victims of immigrants and asylum seekers "who may pass by and pass their way into our communities" (Ahmed, 2004, p. 80). Immigrants and asylum seekers, like terrorists, are identified as potential intruders. Public opinion's fears (e.g., foreigners are a threat; they take jobs away; they increase criminality; they threaten our national identity etc.) are reflected by the media that in turn affect the politics of fear. Thus, the politics of fear sustain the affective conditions that enable antiimmigration and nationalism to flourish, curtail civic liberties, promote attacking everyone who is *different*, and stifle dissent as being *unpatriotic*. I argue, then, that expanding the discourse of fear to include immigration and asylum seeking perpetuates the politics of fear, which follows the establishment of affective (and literal) borders that exclude *the enemy*. The politics of trauma is also sustained, because the global economies of fear strengthen boundary formation; thus, the trauma narratives of immigration and asylum seekers are essentially delegitimized and eventually get lost in the networks of fear. This is most violently revealed, as Ahmed writes, "in the literal deaths of those seeking asylum in containers, deaths that remain unmourned by the very nations who embody the promise of a future for those seeking asylum" (2004, p. 80).

AFFECT AND BOUNDARY FORMATION IN EDUCATION

Insights based on the previous chapters make possible the critical interrogation of affect and the resulting boundary formation in educational contexts. No longer do affects need to be viewed through the narrow prism of individual psychology; rather, they can be theorized in the light of sets of beliefs and practices in their political, social, and cultural contexts. The politics of fear is established not at the level of the individual but rather through the affecting of movement; bodies are (re)socialized as they face others who are perceived as familiar or stranger. This process makes visible the symbolic violence that is exercised upon individuals and communities as a result of hegemonic discourses of fear.

I now want to relate the politics of fear and the educational processes through which fear works to secure the forms of the collective; that is, how the politics of fear makes the world become more fearsome through perpetuating ignorance about what or who it is we fear. This is an important aspect of the affective politics of fear that educators cannot afford to ignore, given the evidence that fear of what is different creates important barriers among children, especially after the events of September 11, 2001 (Horowitz & Wanstrom, 2003; Salaita, 2005; Wohlwend, 2006).

Based on my own empirical work in the United States (e.g., see Boler & Zembylas, 2003; Zembylas & Boler, 2002) and the research presented in several chapters of this book (see also Zembylas & Karahasan, 2006), as well as the analysis conducted by others both in Europe and the United States (e.g., see Ahmed, 2003; Ibrahim, 2007; Kenan, 2005), it seems that the politics of fear have become a dominant motif not only for news and popular culture but also for education. For example, in my ethnographic work with Megan Boler in the United States after the events of 9/11, we described how the processes of fear generation among students were spelled out as patriotic declarations of love and/or anger and resentment against Muslims. In our analysis, we emphasized how students' emotions of patriotism (e.g., anger, fear) provided clues to the constitution of *patriotic subjectivity.* Subjectification was located in a complex set of practices within which the "patriotic self" had been fabricated, and which presupposed and enjoined particular relations with others: from categorizing and labeling *others,* to chanting *patriotic* songs, narrating stories of *courage* and *heroism* found in the media, and analyzing past actions of the nation's heroes found in textbooks. In this manner, frightened students came together, adhering to the nation and distancing themselves from those who did not belong to the nation. Other educators and social scientists also highlighted how narratives of fear about others who were suspected to be linked to terrorist networks led to numerous cases of ethnic or racial profiling, and how the word *terrorist* became associated with anyone who looked Muslim or Arab (Ahmed, 2003; Wohlwend, 2006). In this manner, fear gradually created and reinforced previous affective boundaries that policed the insiders from the outsiders (Salaita, 2005).

Various analyses of pedagogy, ethnic identity, xenophobia, and stereotyping have also been offered in order to highlight these boundaries in Europe. The affective politics of fear in Europe is shown in struggles for defining and protecting European citizenship from the perceived threat by illegal immigrants and asylum seekers (Balibar, 2003; Buonfino, 2004; Kofman, 2005). Various movements of exclusion, strict drawing of external borders, and stigmatization and repression of all those who are deemed as outsiders have created narratives of fear that deserve close and careful analysis by educators. For example, some studies raise concerns about students' feelings of intolerance toward immigrants (Van Peer, 2006); analyses of civics education curricular intent have also shown that different priorities of European countries in relation to national and European citizenship goals create tensions about insiders and outsiders (Ortloff, 2006; Sutherland, 2002).

The evidence from the United States and Europe shows how schools have been caught between reproducing the politics of fear and struggling to find ways to subvert it (Nash, 2005). Thus, exploring the affective politics of fear in relation to the construction of hegemonic discourses and boundary formations becomes increasingly valuable for educators. Insights from the work discussed here show how these boundaries occur through affective processes that are grounded in everyday classroom situations and encounters, including the language that is used in reference to others and the practices for reproducing the apartness with them. It is within this affective terrain that educators need to analyze the implications of practices such as forbidding the Muslim scarf in French schools or raising the national flag as part of school institutional routines. It is worthwhile to examine how the politics of nationalism and its association with such practices may potentially be *mis*-educative in that they are built around the exclusion of others. Discourses of national belonging in schools often include notions of ethnic purity and make rigid distinctions between rigid selves and polluting others (Svašek, 2002). As such, these educational discourses and practices exert *pedagogic violence* (Worsham, 2001), as a result of the way the politics of fear address and educate emotions and cultivate a fearsome affective relation to the rest of the world.

In my most recent ethnographic work, I examine how the politics of fear and hatred are linked in the formation of ethnic belonging among Greek-Cypriot college students (e.g. see chapters 3, 5, 8; Zembylas & Karahasan, 2006). In this work, it is shown how Greek-Cypriot students have learned over the years to fear and even hate members of the other community (Turkish Cypriots) as well as immigrants and asylum seekers. In particular, the role of history curricula and textbooks is analyzed in terms of how students construct a social field in which they are predisposed to enact such negative feelings about the Other. As it has been argued, fear plays a crucial role in the formation of a social or ethnic belonging as apart from others who are perceived as threatening. History textbooks, in particular, are prime candidates for perpetuating the essentialism of such belonging and the fear of the Other (Davies, 2004).

The economies of fear built through history textbooks and curricula contribute to the conservation of existing power relationships by securing prevailing social and emotional norms in relation to the Other. Thus, it is revealed, for example, how particular narratives taught in schools from an early age impose certain affective associations that stick various signs together—that is, Turks/immigrants/asylum seekers are the *fearsome Others* for Greek Cypriots; Turks are fearsome because they have the military power to take over Cyprus, and immigrants and asylum seekers enter Cyprus in increasing numbers and threaten to take jobs away and change the local culture. The economies of fear in schooling are used in the sense of circulation and distribution of particular emotional embodiments through schooling practices and policies. Thus, practices such as commemorating national anniversaries and raising the national flags of Greece and Cyprus in every possible occasion along with discourses of fear that others threaten *our* physical and ethnic survival are highlighted as fundamental mechanisms with which the individual student comes into being through his/her alignment with the collective.

In general, to paraphrase Harding and Pribram (2004), an economy of fear indicates how educators and students give and withhold emotional resources around fear, form social relationships and differences, negotiate meanings, and constitute their subjectivities. The notion of economies of fear is particularly useful in forming a pragmatic understanding of fear and its cultivation through schooling practices and policies, because it emphasizes that fear does not reside within an individual (i.e., a student or a teacher) but is circulated across individuals and between bodies. In addition, this notion shows how fear draws bodies together, making them members of a group united by their fear of other groups.

CONVIVIALITY: AN ETHICS AND POLITICS OF SMALL OPENINGS

When one considers the implications of the affective politics of fear in terms of contributing to the perpetuation of economies of violence and suffering, one is left with the challenge of how to respond ethically, politically, and pedagogically. How can fear be transcended in a viable and skilful way without denying or ignoring its meanings and powers? Navigating the spaces of fear in the context of the contemporary world is arguably a difficult task. However, one comes to a crossroad where the break with past discourses of fear and trauma has to be made inevitable. Instead of performing opposites like "our" trauma against "their" unjust acts by building dubious distinctions of inclusion and exclusion that are too categorical, the task becomes to construct a counterpolitics that is based on the evolving narratives of becoming. To illustrate the possibilities that are available, I want to highlight Gilroy's notion of conviviality. This notion, as I argue, offers worthwhile opportunities for performing what may be called an ethics and politics of *small openings;* that is, possibilities for cultivating individual and collective political consciousness and criticality against fixed and reified identities of the past. Educators, in particular, need to invent pedagogies that account for the processes of affective struggles in community-building as well as the building of local coalitions based on commonalities, without perpetuating past categories (Zembylas, 2007).

In his book *Postcolonial Melancholia* (published in the United Kingdom as *After Empire: Melancholia or Convivial Culture?*), Gilroy (2004) explores the current theoretical tensions and possibilities around the term "multiculturalism." As a way out of the debilitating effects of multiculturalism and the fixed classifications of race, culture, identity, and ethnicity, Gilroy proposes the term "conviviality" to describe "the process of cohabitation and interaction that have made multiculture an ordinary feature of social life in Britain's urban areas and in postcolonial cities elsewhere" (2004, p. xv). Conviviliaty, he explains, is a social pattern in which different groups dwell in close proximity, but where their racial, linguistic, and religious particularities do not add up to discontinuities or insurmountable communication problems. Conviviality acknowledges diversity without cementing it; that is, conflict is not banished, yet conviviality is shown to equip people with means of managing conflict without becoming fearful and hostile—for example, people live on an everyday basis in ways that render race insignificant and inessential.

Conviviality, however, does not signify the absence of racism; instead, it can convey the idea that the means of overcoming racism have evolved. Racism still exists but now it can be analyzed differently. Building on Frantz Fanon's (1980) work, Gilroy analyzes race as a relationship and racism as a discourse with metaphysical capacities. So far, racial discourses of the past were dependent upon fixed racial classifications and ordering. However, conviviality complicates these classifications, including the structured categories of community, difference, and identity, and requires a critical distance from the usual attachments to such categorizations. In fact, to achieve conditions of enabling conviviality, these categorizations must be relinquished and thus accept the risky landscape of movement and becoming. Instead of remaining stuck to the political ontology of race, notes Gilroy, convivial cultures discover that the things that divide or unite them are much more profound: taste, lifestyle, leisure preferences. By making racial differences appear ordinary and banal, Gilroy wants to show that convivial interaction enriches our lives and enhances our solidarities and democracies so that they stop operating in segregated forms.

By extending this thought to include an interrogation of the economies of fear, conviviality inspires us to operate in such forms of living that, even if they are not entirely free of fear, might be better equipped to deal with the implications of fear as matters of ethics and politics, away from narcissistic accounts grounded on fearful claims and denials. Gilroy helps us address the ethics and politics of fear in the contemporary world, without falling into the trap of remaining attached to an endless economy of fear. His vision of "planetary humanism" constitutes a powerful critique of dominant categorizations that cannot escape from the past. Arguably, conviviality is insufficient to prevent lapsing back to such categorizations or moving us beyond fear once and for all and bringing an end to racism and nationalism.

Conviviality is manifested in working to engage with the question of the relation to the Other while thinking past fear, hatred, and resentment. Gilroy suggests that "multicultural ethics and politics could be premised upon an agonistic, planetary humanism capable of comprehending the university of our elemental vulnerability to the wrongs we visit upon each other" (2004, p. 4). This move recognizes the ambivalence of fear, that is, the sense of moral indignation for being treated unjust by the Other but also the feeling of shame at recalling one's own horrible acts against the Other. The ethics and politics of conviviality do not work against the ambivalence of fear, but work within it and articulate a critical discourse from which to launch a renewal; that is, a restructuring of affective space in such a way that new affective connections with others are made possible. As each community is trapped in its egotistic mentalities, conviviality interrupts the endless cycle of selfishness and makes visible the suffering of others as well (see also Metz, 1972, 1980, 1999; chapter 8).

IMPLICATIONS FOR EDUCATIONAL POLICY, THEORY, AND PRACTICE

An interrogation of the affective politics of fear allows educators to critically reconsider the ways in which educational policies and practices lead to particular affective economies within a pedagogic field (e.g., a classroom or a school). For instance, in what ways do such policies and practices encourage or inhibit open-minded, tolerant, and democratic dispositions among students? In the context of feminist, critical, and poststructuralist conceptualizations of power and social relations, and emotion discourses and practices, educators can investigate how a classroom community establishes power relations within historically specific configurations of affective networks. For example, an inquiry may be launched on the historically changing forms of social control that educational policies and practices employ to maintain emotional structures that are in their own interests such as the notion of the *disciplined child* (e.g., see Popkewitz & Brennan, 1998). Such policies and practices are facilitated by the modes of emotional engagement and are perpetuated through the cultivation of particular embodiments—such as rituals and collective behavior that contribute to the formation of inclusions/exclusions. In the final part of this chapter, I will show that an interrogation of the politics of fear in education can create openings that would allow for the possibility of critical resistance to the local or global economies of fear, even if resistance rarely succeeds in bringing about genuine transformation (Hoy, 2004).

Interrogating the politics of fear in education makes possible an analysis of the affective networks of fear within particular sociopolitical fields; these networks involve movements of emotional capital of fear. For example, fear of Muslims after 9/11 or of immigrants in Europe is not inherent within a particular body of a student but rather is an effect of the affective encounters within a *community;* students who belong to the dominant group move toward or away from marginalized others. As emphasized in previous chapters, the alignment of some bodies with some others and against others takes place through the affecting of movement; bodies move away from others or close to some others who are recognized as fearsome or loved, respectively (Ahmed, 2004). That is, the emotional dispositions of fear *do* things, because the marginalized others are transformed into the fearsome others who threaten to harm us. Fear is easily distributed through everyday practices in schools; practices that come to signify the danger of impurity from mixing with *them*. Needless to say, an affective economy of fear is not a blanket that equally covers each member of the group associated with it; each group member will contribute differently to the formation of this economy and thus will experience it in different ways (Barbalet, 1998). In other words, as it is shown below, there are always *small openings* for subverting this economy of fear.

The above arguments urge educators to consider how the cultivation of fear of those who are *different*—through curricula, textbooks, and everyday school practices—organizes the social and bodily space in a pedagogic field by creating powerful affective borders. Consequently, the task of any pedagogy that is critical of these affective borders among individuals and groups is to identify the practices, strategies, and spaces in which affective transformation might be possible. In other words, if our commitment is to individual and social change, as Worsham (2001) notes—that is, change that would finally dissolve the normative tendencies of the affective politics of fear in educational policies and practices—then the work of decolonization must occur at the affective level, not only to reconstitute the emotional life of students and teachers but also to restructure the affective field of pedagogy itself. Our most "urgent political and pedagogical tasks," asserts Worsham, become then "the fundamental re-education of the emotion" (2001, p. 233).

The first implication of the account sketched here is to analyze the modes through which fear is authorized by, implied, and embodied in pedagogies—and thus not misrecognize fear as a problem of an individual. Embodied practices emerge from specific teaching and learning spaces in which bodies, psyche, and memory interact (Probyn, 2004). This interaction produces a bodily knowledge that has the potential either to perpetuate or to interrupt the stigma of evil that is granted to those who are different. Consequently, to take seriously the affective politics of fear in pedagogies is to explore how such modes of fear are part of the efforts to articulate a relationship between the self and the Other (Giroux, 1997). That is, analysis of the politics of fear involves the interrogation of the ways in which explicit mobilization of fear produces social relations. Educators need to question, for example, whether these social relations simply tolerate otherness—toleration as that which implies something to get over—or really welcome the Other as the link through which worthwhile networks might emerge (Gordon, 1999). The question, writes McCormack (2003), "is not only 'how far can we care,' but also becomes one of cultivating a commitment to those relations that may increase the intensity of attachment and connectivity" (p. 503).

The second implication, then, is that inasmuch as a classroom or school emotional climate emphasizes the ways in which schooling perpetuates the hegemonic discourses of fear, educators need to examine the individual and community spaces that may be constituted to disrupt the modes through which fear is embodied. A task such as this requires the transformation of the teachers' and students' dispositions within an affective network of relations (e.g., conviviality). Such a transformation emerges out of the very practices and strategies in which relations operate, that is, by changing the practices enacted in a classroom to produce alterative emotional capital. For example, engaging in inclusive practices rather than exclusive ones that further divisions is an important step toward reestablishing closeness between bodies whose proximity is read in an oppositional manner; the literature on inclusive, antiracist, peace and reconciliation education is full of practical examples of how to do so (Zembylas, 2007).

The disruption of normative embodied practices through the production and circulation of different emotional capital is certainly not an easy task for educators. The dispositions of the hegemonic discourses of fear are not easily suspended by a simple effort or will. Affective communities embedded in fear are not easily undone through a liberatory educational program that simply tackles ignorance about the Other. The issue, as Bourdieu asserts, is not "a simple 'conversion of minds' . . . produced by rational preaching and education" (2000, p. 180), because passions are rooted in emotional dispositions. Besides, there is considerable research suggesting that a change of heart does not follow, naturally or simply, from a change of mind (Neu in Worhsam, 2001; Zuzovsky & Yakir, 1999). What is most often required is a transformation of the very conditions, emotional and otherwise, of the production and re-production of the economies of fear.

In other words, what is required in a classroom or a school community is initiating a different *emotional climate* that is not stuck in the politics of fear and trauma (cf. de Rivera & Páez, 2007; de Rivera, Kurrien & Olsen, 2007) but paying attention to particular practices that promote conviviality, solidarity, empathy, and reconciliation. The usual focus of many multicultural education programs on tackling one's ignorance about others or acquiring better social skills of communication tend to overlook the affective aspects of transformation. In such programs, the emotions are seen to matter insofar as they are derivative aspects of cognitive processes (Emirbayer & Goldberg, 2005). Such reductionism leads to the lack of considering the alignment between the individual student and the collective. However, describing emotional dispositions as embodied within a specific social context allows a connection between individual psychic patterns and the broadest accounts of transformation; the affective community is indeed the very site of the capacity to effect change. Embracing affect as a necessary part of the struggle embedded in a pedagogic field can make a difference by moving teachers and students emotionally, ethically, and politically toward becoming critical witnesses, not spectators of the violence exerted by the politics of fear (Zembylas, 2006b). For example, students may be asked to map the terrain of their own community/family history (through drawings, paintings, written work etc.) and collect testimonies of discrimination against others (e.g., those who have a different color or ethnic background) as foundations for a social ontology of resistance. Becoming critical witnesses implies that students interrogate the affective ways with which individuals are involved in social structures.

The importance of analyzing the affective politics of fear in education is the recognition that likeness and unlikeness of us and them are parts of the work of emotion. Therefore, the only way of undoing these affective boundaries is to open the community to the marginalized ones and reeducate the emotions; this constitutes a significant political and ethical act of re-forming the affective connections with others. Bennett (2001) discusses the significance of such affective connections that emerge through the development of the emotional dimensions of sensibility. For Bennett, affective engagement—for example, feeling the repercussions of the politics of fear on an everyday basis as exclusion, shame, or guilt—is an essential component of an ethical life. On Bennett's view, contemporary critical theories seem to adopt an attitude of emotional detachment when it comes to ethics—a position that leads to feelings of alienation, bitterness, and hopelessness. On the contrary, Bennett highlights the *force* of affectivity as a source that might propel ethical generosity. For her, the experience of this affective force arises from an encounter or a crossing, that is, social networks among people.

An economy of fear serves both to maintain social and political belongings and to be the pivot of collective behavior or action. Reeducating the emotions, then, arguably represents an effort to change economies of fear, through seeking to expose the relationship between the individual and the social structures, between the legitimate and the illegitimate, and to open for examination the symbolic violence implied in teaching and learning (Worsham, 2001). The ideas discussed here encourage the exploration of the ways in which mechanisms of exclusion address and educate fear and establish fearsome relations with others. Educators and learners need to look beyond dominant curricula and the reproduction of existing knowledge and embodied practices and consider the implications from an account that interrogates the affective politics of fear. Pedagogies that wish to disrupt the hegemonic politics of fear—such as cultivating critical witnessing, opening the community to others, and inspiring practices of empathy and reconciliation—must teach critical resistance. Pedagogies that wish to disrupt the hegemonic politics of fear in the contemporary schooling of the United States and Europe are essentially decolonizing pedagogies in the sense of offering alterative practices and discourses that broaden hope and extend the range of possibilities for conviviality.

Chapter 5

The Politics of Shame in Intercultural Education

How do shame and pride politics police the terms of belonging to a nation-state and work to bring a nation together as a *felt* community? How does the ideological role of schools seek to eradicate shame via an erasure of certain histories and the sanitization of past wounds inflicted on marginalized others? Can intercultural education help to heal past wounds through a different understanding of shame, a potentially *productive* one? In contemporary debates on the place of shame in political life, there seems to be a clear boundary between shame and pride; shame is considered a *destructive* emotion that needs to be avoided while pride is promoted as a salutary emotion that serves as a legitimate force of nation building (Tarnopolsky, 2004). This dichotomy between shame and pride, argues Tarnopolsky, is not only oversimplistic but also arises out of an inadequate understanding of shame and its place in politics. Similarly, the place of shame in educational discourses is not only undertheorized and (national) pride receives an extraordinary attention compared to shame, but also there are efforts to completely erase discussions on shame via an erasure of histories of marginalized others (Fortier, 2005).

It is well documented, for example, especially in areas of conflict that educational practices are used to create nationalist subjects and eliminate discussions on shameful acts of the past such as atrocities and violations of human rights at the expense of others (Feuerverger, 2001; Salomon & Nevo, 2002; Tawil, 1997). Systematic efforts are undertaken in curriculum development to implore students to remember the pride brought to a nation-state by the leaders and warriors who defended its lands and values. A careful look at the curricula in conflict areas such as the Balkans, Israel, or Cyprus offers evidence on how past glories are highlighted while shameful actions are erased (Cole, 2007; Davies, 2004). Students in those areas are repeatedly reminded of what it means to belong to an ethnic or racial group by reasserting particular values, principles of patriotic responsibility, and moral conceptions of right and wrong. It may well be argued that political struggles on national identity are struggles for *recognition*—one of the foundational issues in multicultural debates over the last 20 years (see Fraser & Honneth, 2003). The assumption here is that *positive* recognition is achieved only through pride politics; there is no room for considering shame in these debates. Even in cases that shame politics is considered, efforts are made to hide its implications and *sanitize* the stories that are told. To struggle for recognition is to struggle for a picture untarnished, as Bingham (2006) rightly points out.

What is striking, however, in this rhetoric, is not only how shame provides indication to the structural dimension of both nonrecognition and misrecognition of subordinate groups, but also how it constitutes a powerful form of nation building that further polarizes cultural groups within a nation (Ahmed, 2004; Probyn, 2005). This polarization is extended by educational efforts to police the boundaries of belonging and entitlement to citizenry through legitimizing and delegitimizing specific practices and behaviors. Unavoidably, then, some groups within a nation may be excluded as a result of both institutional structures and misrecognition or nonrecognition of their histories. A classic example of this exclusion is the group of Aborigines in Australia. In the last few years, much evidence is coming up on the systematic political efforts by Australian governments over the years to play the card of national pride in order to eradicate shame for the trauma caused to Aborigines (Probyn, 2005). In these efforts, the Aborigines have been either misrecognized or not recognized at all, argues Probyn, because their stories have been erased and a more sanitized history has been promoted. Again, the foundational assumption made in efforts to promote a more sanitized history has been that shame is harmful for the nation (which nation?) and thus references to it ought to be eliminated.

However, if one wants to transcend this dichotomy between shame-as-bad and pride-as-good and the relation between shame and pride is not staged as one of simple opposition, then, we may begin to explore that shame can also be productive—in so far as it adds something to social and cultural encounters, providing new understandings of how different cultural groups inhabit society. For example, it is not impossible that recognition of shame for the injustices committed against marginalized others may work to reconcile a nation to itself by coming to terms with its own past (Ahmed, 2004). Or, in providing opportunities for a nation to "feel bad" about its past wrongdoings, shame may allow a nation to consider how people change as a result of self-evaluation (Probyn, 2005). In fact, there are specific examples drawn from reconciliation and antiracist practices (e.g., see Berlak, 1999, 2004; Boler & Zembylas, 2003; Cherry, 2004; Doxtader & Villa-Vicencio, 2003; Zembylas, 2006b, 2007) suggesting that exploring the use of shame can indeed goad action and inspire rapprochement, rather than perpetuate existing dichotomies between self and Other.[1]

In this chapter, I explore the openings that are created for a productive politics of shame in the context of discussions surrounding intercultural education via tracing how different kinds of shame are evoked or rejected in collective imagination—both in politics and in education. My goal is to show that a rehabilitation of the political and ethical value of shame challenges the opposition between self and Other and that this act can be beneficial, particularly in educational settings. The misrecognition of shame as something destructive both in political and in educational debates fails to capture two important insights, as Tarnopolsky (2004) suggests: first, that shaming creates an opening that is necessary for self-criticism, self-reflection, and ethical, political, and educational deliberation; and second, that shaming under some circumstances may encourage the possibility of solidarity and conviviliaty, an invitation around which both the group-shaming and the group-being-shamed might *meet*. Thus, an effort to transcend the usual dichotomies around the pride/shame debate both in political and in educational discourses will tackle the cultural, political, and educational uses of shame and help educators gain a deeper understanding of the meaning and implications of shame in intercultural education. Such a renewed understanding adds a significant dimension to contemporary discussions about recognition.

THE POLITICS OF SHAME IN CONTEMPORARY DEBATES: A BRIEF GENEALOGY

In recent years, there has been a surge of interest in the emotion of shame across a wide variety of disciplines (Probyn, 2005; Scheff,

1994; Scheff & Retzinger, 1991; Sedwick & Sedwick, 1995; Williams, 1993). Scheff and Retzinger (1991) consider shame as a master emotion because of its part in nearly all daily acts—individual and collective ones—comprising humiliation, discomfort, rejection, failure, and insecurity. Scheff (1994) puts this idea even more strongly when he asserts that shame and pride may be the most powerful forces in the human world. In Western societies, shame has been defined in such an extremely negative way that it is often denied and goes unacknowledged; thus, in their extensive work in this area, Scheff and Retzinger insist that if shame is not appropriately understood and acknowledged, it results in aggression, resentment, and intractable conflicts (Scheff, 1994; Scheff & Retzinger, 1991). However, if shame is acknowledged, according to them, it can become the origin of positive things, despite involving extreme emotional pain and social disgrace.

Shame is essentially feeling bad about oneself before others, therefore, it is a sign of severed social bond (Scheff & Retzinger, 1991). Consequently, shame needs to be understood as a social emotion, one that is not only considered intrapsychic but also that circulates, attaches to, and intensifies the meaning and feeling of identity (Sedwick, 2003). That is, shame is about one's entire self as this is constructed within a certain social, cultural, and political context; thus, when shamed, one's sense of belonging is negated. This relationality of shame is crucial in identifying the political and ideological processes with which one is socialized to experience (or not) shame (and pride) for one's in-group actions. In other words, although the jump from shame-as-individual to something collective is not unproblematic, there is significant evidence about the function of shame as relational in ways that make individual manifestations directly relevant to collective ones (Scheff, 1994; Scheff & Retzinger, 1991).

Another challenging issue concerns the considerable disagreement about defining and understanding the uses of shame and its differences from guilt. Neither of them is construed as a one-dimensional construct; instead, they include families of emotions—for example, anger, fear, sadness, and so on. Shame and guilt are almost invariably paired together, although there are important differences between them (Forrest, 2006). These differences can be identified at various levels. At a chronological level, it is argued that in the course of moral development shame should yield to guilt—a five-year old, for instance, is far too young for guilt (Woodward, 2000). At a historical level, shame is regarded as the more primitive emotion, one that in the normal course of things should be replaced by the more complex and modern emotion of guilt (Scheff & Retzinger, 1991). This notion is based on Freud's theorization that guilt indicates the internalization of values as opposed to shame, which shows the external disapproval by others; thus, ancient Greek culture is motivated by shame to avoid disgrace, whereas modern Western civilization internalizes convictions of right and wrong and socializes people to feel guilt (Scheff, 1994). There is, of course, the counterproposition by Elias (1978) that there has been no decrease in shame but rather an increase in undercover shame—that is, there is now more shame than earlier, but much less awareness of it. Finally, at a philosophical level it is suggested that one who feels guilt acknowledges that he or she made a specific error; when one feels ashamed, his or her entire being is attacked (Forrest, 2006).

One of the striking implications of shame is how it works to reform affective and political spaces—as bodies turn away from some and closer to others—to silence or isolate certain cultural groups. These spaces operate through a rejection of one group as defeated, exposed, and unethical while the other group comes through in some way as superior. There is clearly a discrepancy in how one group is viewed by itself as compared to how this group is viewed by others (Scheff, 1994). In either case, the Other is experienced in some way as unlike the self and the group in which this self belongs—for example, the expression "*they* are different from *us*" is characteristic of this perception. Particularly in conflict areas, pride in *our* history and *our* nation is repeatedly rehearsed by way of an image that knows no shame; shame is left for others (Volkan, 1997). This denial of any shame is simultaneously a self-affirmation for the collective body and a declaration of its self-pride (Probyn, 2005).

The refusal of any shame—for instance, as a result of a nation's past wrongdoings such as in the example of how Aborigines have been treated in Australia—is also a refusal to make shame *ours* in the first place (Fortier, 2005). That is, it is a refusal to consider that shame and dignity can be intertwined as a result of admitting past wrongdoings and acknowledging the commonalities between us and others. According to Fortier, this refusal prevents a unique opportunity for creating new social and political alliances because shame is rejected as something *we* do not have—and should not have—in the first place. Shame is projected on to those who are considered as different from *us* (e.g., those who are less patriotic; those who do not belong to the nation; and those who feel more committed to their own cultural community instead of the nation). Thus, others are essentially blamed for their inability to deal with shame. Invested in the process of eradicating shame from the collective body is a process of preserving the polarities between us and others. This form of politics of shame works by further stigmatizing certain groups and individuals. The misrecognition of the terrain of the political and social as one that sets clear boundaries between shame and pride fails to do justice to the complexities of the different ways that shame may work.

And yet, as Silvan Tomkins emphasizes, shame operates only when there is *interest* involved (Sedwick & Sedwick, 1995). Drawing on Tomkins's link between shame and interest, Probyn (2005) and Sedwick (2003) provide a different account of shame. They argue that shame represents lines of connection between people and functions as amplification—that is, it makes us care about things. If we feel shame, as Ahmed (2004) asserts too, it is because we have failed to approximate a social and political ideal about the Other; thus, shame is the loss of indifference. Consequently, it may be argued that shame sensitizes us to a vast variety of actions—actions that have either brought shame upon us or caused others shame.

But shame does more than sensitizing us; it also proposes a sensibility at once practical and ethical, because the appropriate reaction to one's own shame is a type of self-transformation, as Redding (1999) puts it. Probyn writes: "Shame makes us question what we are feeling, the nature of the loss of interest, and fundamentally . . . who we are, as a reevaluation of the self" (2005, p. 64). Thus, shame is important "to discussions and debates about how to deal with pasts that could be called shameful; and to visions of life curtailed by the idea that there is something intrinsically wrong with feeling shame" (Probyn, 2005, p. xiii). Such a view of shame makes possible a space of solidarity that transcends feelings of resentment and anger invested in past collective histories of trauma and violence. It may be said that a political and ethical account that subverts the notion that shame is deeply shameful has important implications for the kinds of solidarities that may eventually be built. As such, shame promises an interruption of traditional identifications (e.g., with *a* nation, *a* religion) and offers ethical and political leverage, "one that . . . has its own, powerfully productive and powerfully metamorphic possibilities" (Sedwick, 2003, p. 65). As Sedwick suggests, a move away from the notion of shame as disapproval can offer new ways of avoiding the fixation on the dichotomy between shame-as-negative and pride-as-positive—that is, the goal is a politics of shame that does not rest on an essentialist conception of shame.

National Shame in a Multicultural Society

But what exactly does it mean for a nation to "admit to" feeling shame particularly in a multicultural society comprised of various groups that often have grievances against one another? Can this admission help rectify past wounds and claim a new space for a postnational identity? Would it be appropriate to engage in such discussions in the context of intercultural education programs?

Two examples may help contextualize these questions in a wider political arena: The first example is the *Bringing Them Home* report (Human Rights and Equal Opportunity Commission, 1997) and the second example is Sorry Books—both examples are situated in the Australian context (see Ahmed, 2004; Probyn, 2005). The *Bringing Them Home* report was the result of an inquiry that provided evidence about the horrible experiences of Aboriginal people as a result of the Australian government's policies that removed in force thousands of Aboriginal children from their families (known as the Stolen Generations). In the report, the Australian nation recognizes the wrongdoings of the past and acknowledges that the government policies are aimed at the genocidal eradication of the Aboriginal people. In particular, in the introduction of the *Bringing Them Home* report, the governor-general of Australia is cited in a statement he made in August 1996:

> It should, I think, be apparent to all well-meaning people that true reconciliation between the Australian nation and its indigenous peoples is not achievable in the absence of acknowledgement by the nation of the wrongfulness of the past dispossession, oppression and degradation of the Aboriginal peoples. That is not to say that individual Australians who had no part in what was done in the past should feel or acknowledge personal guilt. It is simply to assert our identity as a nation and the basic fact that *national shame*, as well as national pride, can and should exist in relation to past acts and omissions, at least when done or made in the name of the community or with the authority of the government. (Human Rights and Equal Opportunity Commission, 1997, p. 3, emphasis added)

In the report, the legacy of the forced removal of children is described as "shameful" (1997, p. 390); it is therefore suggested that the identity of the nation has to be redefined through shame.

Ahmed (2004) contends that the readers of the report are in a double and paradoxical position. On the one hand, they are asked to acknowledge shame as their own shame (i.e., national shame), and on the other hand, they are called to be witnesses of the pain of others and the injustice committed against them. This double position draws out the ambivalent yet self-reflexive aspects of shame as something that might be expressed about some issues but not others. Shame compels an immediate reassessment of self, individually and collectively (the national self): Should I/we feel ashamed? Why/Why not? Can national shame offer insights into what an apology—or a refusal to apologize—may achieve? Shame in this way, asserts Probyn (2005), can be positive in its self-evaluative role and may even be self-transforming. But unlike Probyn, Ahmed focuses on the role of witnessing in the ways national shame redefines the sense of belonging. As she writes, the "narrative of recovery" serves as the "re-covering of the nation" (2004, p. 112). Against a view of shame as shameful and paralyzing, Probyn and Ahmed argue for a notion of national shame as a mechanism for reconciliation, in which acknowledging past wrongdoings provides the ground for claiming a new politics of identity.

In the second example, Sorry Books may also be examined as an activity that offers opportunities for witnessing the Aboriginal people's suffering in Australia. Sorry Books refer to the writings of Australians (mostly nonindigenous) in which condolence and sorrow are expressed for past injustices committed against Aboriginal Australians; these books exist in both paper and electronic forms. Sorry Books, explains Ahmed, function "as a demand to and for the nation to appear ashamed, and to speak the shame on behalf of Australians" (2004, p. 110). This demand also expresses the need that the nation moves beyond the past. The denial of shame on the part of the official government is seen by the supporters of reconciliation as a process of eradicating shame from the collective body via an erasure of past wrongdoings. Thus, it may be argued that this denial of shame actually denies Australia a unique opportunity to redefine its identity in a multicultural society.[2]

Both examples suggest that witnessing past wrongdoings does not necessarily paralyze a nation or threaten its unity; on the contrary, witnessing allows individuals and groups to empathize with others who have suffered or still suffer (Oliver, 2001). Although shame is acknowledged and past brutalities are recognized, the nation does not have to remain fixed on either denying shame or highlighting pride. A multicultural society can be united under the ideal of solidarity and empathy and transcend polarities such as that of shame/pride that may lead groups further apart.

Similarly, the publication of the Parekh Report in Britain (Runnymede Trust, 2000) is another example that brings the issues of national shame and national pride in the middle of intense debates about multiculturalism and shows that the simple distinctions between shame and pride are problematic. The Parekh Report posed a fundamental question by asking whether it was possible to reimagine Britain as a postnation in a multicultural way. The media attacked fiercely the report's suggestion that Britishness carries racial connotation, which was misread as an accusation of Britishness as inherently racist (Fortier, 2005). The controversy, according to Fortier, was overshadowed by declarations of national pride for being British and denials for accusations of shame. Fortier analyzes how the resort to the discourse of national pride and national shame is linked to the process of collective identity formation and writes, "The pride/shame debate reveals how the politics of pride seek to eradicate shame via an erasure of certain histories, and to sanitize Britishness under a veneer of tolerance" (2005, p. 562).

Discourses that remain fixed on the shame/pride polarity fail to recognize the place of racism, oppression, and wrongdoing—and thus feelings of shame for those practices—in the historical process of how a modern nation-state has been built (Probyn, 2005). The desire for national pride and the repulsion for shame in almost all modern nation-states since the nineteenth century has become a major mechanism of self-affirmation (Barkan, 2000). Yet this desire to feel pride often involves the erasure of past histories of violence, trauma, and suffering against those who are marginalized. An important opening to gain a renewed sense of pride through expressing sorrow and questioning existing social structures is missed. Instead, any attempt by marginalized groups to raise such issues is condemned as unpatriotic and shameful, and is responded to in anger against those who "shame the nation," as if "the maintenance of a guilt-free national story" (Fortier, 2005, p. 566) will foreclose any responsibility for the past.

Undoubtedly, political as well as educational efforts to subvert the shame/pride polarity include discomfort and ambivalence but a politics of shame does not have to be normalizing, stigmatizing, and destructive. On the other hand, there is also the danger that acknowledging shame may be done superficially or in a moralistic manner for the purpose of reinstituting national pride without really accepting responsibility for the past. Saying sorry, for example, does not necessarily subvert shame to pride and it is certainly not an indication of assuming responsibility or building solidarity with marginalized others (Spelman, 1997). Solidarity is possible only when the politics of shame foregrounds the historical conditions in which trauma is produced, inflicted, and experienced, and it demands that all of us place ourselves in relation to that history (LaCapra, 2001).

Instead of finding ways to quickly eradicate shame, we must try to find a kind of dignity in shame that will draw us together in the recognition of our indignities (Tarnopolsky, 2004). In the remainder of this chapter, I utilize the ideas sketched here to read how some examples from intercultural education may indeed challenge an oversimplistic opposition between national shame and pride. My aim is to consider how a productive politics of shame in intercultural education offers a deeper understanding of otherness, identity, and past historical trauma.

Toward a Productive Politics of Shame in Intercultural Education

I will begin by describing two examples that articulate how the politics of shame play out in a classroom, and how complexities around issues of trauma and shame related to nationalism or racism actually operate for some students. The first example is taken from my ongoing ethnographic research on trauma and witnessing in Cyprus. As I have pointed out in chapter 3, each side in Cyprus—that is, Greek Cypriots and Turkish Cypriots—constructs narratives in which the Other is portrayed in negative ways, but the common themes focus on the national pride that each group feels for its glorious past and the victories inflicted on the other side. In the official historical narratives taught in schools, there is nowhere to be found any references to shame as a result of the atrocities that each community has committed against the Other. In other words, the oppositional relation between pride and shame is implicit both in political and in educational discourses (Bryant, 2004).

As an instructor who teaches preservice and in-service teachers (Greek Cypriots) in this troubled part of the world, I try several pedagogical interventions that aim at exploring the affective implications of past trauma narratives on peace and reconciliation efforts between the two communities in Cyprus. In one such intervention, my preservice teachers were asked to collect trauma narratives about the intercommunal conflict in Cyprus. Not surprisingly, all of my students collected trauma stories only from their own community. After several weeks of collecting and analyzing these stories, the issue of national pride was brought up. First, I posed the question, "What do you think you learned from collecting these stories?" Several students suggested that they had not realized the extent to which past trauma and suffering affected members of their community on an everyday basis. Then, references to national pride monopolized our discussion: Both the majority of the individuals who had been interviewed as well as many of my students referred to the pride they felt for belonging to "a great nation" and their sorrow for the trauma their community had suffered as a result of the atrocities committed by the other community. Although, there were a few references here and there about crimes committed by Greek Cypriots against Turkish Cypriots as well, the discussion was dominated by the crimes committed by "the Turks" (no distinction was made between Turks and Turkish Cypriots) against "us," there was not a single reference to any feelings of shame or regret for the pain "we" inflicted on the other side.

Next, I invited a Turkish-Cypriot friend, Hakim, to speak to my students about the *other* point of view. Hakim narrated his life story and the pain and suffering experienced by his family in their struggle to survive from the violence and atrocities committed by the Greek Cypriots. Many of my students were surprised to hear his story and confessed that it was the first time they had ever met a Turkish Cypriot. They told Hakim about their own preconceptions concerning the Turkish Cypriots and many of them soon admitted that stereotypes had played a negative role in the communication between the two communities. In their journal reports written the day after Hakim's visit, my students analyzed various aspects of how nationalistic education had contributed to having negative feelings about the other community. Several students acknowledged that they felt confused and mentioned feelings of shame. One student wrote,

> I don't know how to write this. . . . This experience was shocking to me. It's the closest I ever came to feeling shame about my community and the atrocities it committed against the other community. All these years I grew up feeling pride about my community's past. Now I am confused. . . . But the discussion we had after Hakim left helped me realize that shame should not paralyze us. On the contrary, we should try to learn the most out of such feelings. I have to admit that meeting Hakim was an extremely valuable experience that taught me a lot. . . . Perhaps the most important lesson I learned was focusing on how each community can acknowledge the other's pain and suffering with dignity; we should not let negative feelings derail our efforts for reconciliation.

The opportunity offered to my students to critically approach the notions of national shame and pride set the scene to begin recognizing nationalism and even speaking against it. However, I doubt that this opportunity helped them develop a deep awareness of nationalism and its complexities around issues of shame and pride. As I pointed out in earlier chapters, the process of critical self-reflection on these issues is painful, complex, and long; yet the ideas to which these students have been exposed offer a richer understanding about the dynamics behind affective attachments to collective identity.

The second example is based on Boler's (1999) notion of *pedagogy of discomfort* and Berlak's (2004) account of her efforts to sensitize her own preservice teachers to issues of social justice. In her analysis, Berlak focuses on a particular incident in which a guest speaker's expression of anger in response to racism aroused feelings of shock, surprise, and discomfort in her students. In this incident, an African American elementary school teacher (named Sekani) visited Berlak's teacher education class to give a presentation about teaching African American children. To provoke students (mostly of European descent) to rethink their assumptions about classroom management in classrooms populated by poor black and Asian students, Sekani presented a role-play situation. This role-play sparked an intense exchange of angry responses between Sekani and a white student.

After the encounter, Berlak describes students' immediate as well as written responses focusing on the denials and affirmations of Sekani's views. Berlak uses Boler's (1999) distinction to discuss Sekani's *moral anger* and many of the students' *defensive anger*. Sekani's moral anger was the anger at socially induced suffering and people's lack of sensitivity about racism, whereas the students' defensive anger was a response to shame for being blamed for injustices. A closer look at the incident and its aftermath help to see the beginnings of a profound transformation. Although it is uncertain to say how successful this change was at the end of the day, the encounter with the Other's point of view seemed to help many students acknowledge how hard it was to question cherished beliefs. Not surprisingly, then, the invitation to question such beliefs was not something that all students readily accepted. Many students felt that Sekani was pointing a finger at them and thus their comments reflected the discomfort encountered when attempting to realize her feelings of suffering and their potential involvement in racism. But as Berlak's analysis suggests, some students were not paralyzed by initial feelings of shame; rather, they utilized the opportunity of experiencing discomfort to take a risk and begin to get out of the binary trap that they are either innocent or guilty. Learning to hear the other's feelings and empathize with those feelings helped these students to acknowledge not only one's own pain but also the pain of others and, therefore, re-cover from binary traps.

As both examples show, the vocabulary of shame is closely linked to particular affective investments in social and political space. In both cases, the moments of witnessing shame consists of discomfort, confusion, and ambivalence, but it does not have to end up being a stigmatizing experience. Therefore, to create social and political spaces of solidarity in the classroom requires taking into account these affective investments. Getting stuck in feelings of guilt and shame or even refusing to acknowledge any shame for past wrongdoings may eventually perpetuate the separation between us and them. Witnessing shame in the classroom, however, is neither about making students feel bad about their ancestors' actions nor about telling them moralistic tales of how to apologize for things that they or others have (not) done. Witnessing shame, as the above examples indicate, is a form of pedagogy that can motivate students and teachers to try and recapture the positive affective connections that past trauma has impeded.

Some may argue that the notion of shame is inappropriate for exploration in schools. However, there is every reason to believe that witnessing shameful acts may indeed create openings for affective connections that are concerned with the present and the future, as a means for initiating meaningful changes. Both examples discussed here suggest that envisioning spaces of respect and tolerance and engaging in acts of affective connectivity—for example, acts of compassion, sociality, and dignity—are precisely what a productive politics of shame in intercultural education may offer to teachers and students. Unraveling the political aspects of shame—both as analytic tools and as points of departure for cultivating individual and collective political consciousness and self-reflection—creates possibilities for enriching teachers' and students' perspectives about the dynamics of affective relations in the political landscape of the classroom. While my argument questions any politics based on notions of liberation as a source of personal/political transformation, it should not be interpreted as a utopian politics; this effort is grounded in the pragmatic prospects of renewing affective relations via community-building in the classroom.

A conceptualization of the pedagogical possibilities that are created in view of the examples and issues analyzed in this chapter enables new directions for a radical reinterpretation of shame, especially in school subjects such as social studies and literature. This reinterpretation happens in terms of the affective connections and spaces in which such possibilities are embedded. In part, these pedagogical possibilities arise from the ambivalence and open-endedness that accompanies a pedagogy that encourages interrogation of issues around shame and pride. Such a pedagogy does not settle issues in any comfortable way and does not seek the settlement of all tensions around an all-encompassing perspective. But through a consideration of the politics of shame, educators and students are enabled to revisit some of the affective implications around multiculturalism, collective identity, and historical trauma. Moving through emotions such as shame, pride, prejudice, ambivalence, and discomfort, and creating powerful connecting landscapes in the classroom can offer transformative opportunities that may shift the emotional boundaries of possibility toward critical hope (Zembylas, 2007).

The preceding analysis also offers a number of useful perspectives that lay the groundwork for an argument about the significance of developing a productive politics of shame in education. First, any attention to shame needs to be placed within a particular context of disciplinary practices and techniques, if it is to be an effective force of affective capacities. In other words, the witnessing of shame as a departure point of problematizing nationalism or racism emphasizes the need to identify how discourses and practices are embodied in the day-to-day routines of school life and what possibilities are opened for interrupting shame/pride polarities. This kind of pedagogy also suggests developing a mode of critique that comprehends the affective economies of schooling as well as their effects on students' and teachers' lives. Students and teachers bring different emotional histories with them to school; these histories are embedded in a wider context of sociopolitical forces, needs, and interests. Developing a pedagogy that examines the politics of shame would mean developing affective connections and social relations that focus on critical interrogation of past histories and experiences. Consequently, intercultural education may be delineated as the space in which victims and perpetrators of past traumas and injustices "can find a way of living [. . .] that feels better through the process of speaking about the past, and through exposing the wounds that get concealed by the 'truths' of a certain history" (Ahmed, 2004, p. 201).

Second, the testing ground for these ideas is the pedagogical efforts to understand the emotional damages of racism or nationalism and their ethical and political implications. Witnessing shame as an affective practice is then an ethical and political project and carries with it a responsibility to explore affects such as dignity, respect, and compassion and to promote their enactments in the community (Oliver, 2001). *Communities of memory* is a good example of a communal structure through which people struggle with the possibility of witnessing and create affective economies that may challenge a community's past (Simon & Eppert, 1997). The contexts of communities of memory, according to Simon and Eppert, include school and university classrooms, as well as informal groups and intimate gatherings. The practice of witnessing within a community of memory involves an active involvement of its members with the affective politics of past memories. This implies that affective relations among the members of this community—for example, feeling for the other's suffering (compassion) or feeling bad for causing another's suffering (shame)—are productive modes of social interaction in which examining testimonies can work round emotional impasses and generate trust and connectivity.

Rak (2003) raises questions whether creating discomforting conditions and bad feelings is an ethically responsible way to teach. She suggests that there are other ways to present, teach, and discuss testimonies in the classroom, without necessarily creating discomfort. Whether or not teachers create discomforting conditions, it is important to acknowledge that the act of witnessing shame opens up a much-needed affective space in the classroom to examine the ways with which traumatized experiences can be witnessed respectfully and thoughtfully (Zembylas, 2006b). Besides, there are no *safe* classroom spaces, if one considers that conditions of power and privilege always operate in them (Boler & Zembylas, 2003). Therefore, the focus, I believe, needs to be on what is required of teachers to develop a critical stance of bearing witness to shame.

In general, the challenge to create pedagogical spaces for witnessing shame is twofold. First, there is the challenge of developing pedagogical practices that analyze and critique the ways in which bearing witness to shame and pride encourages feelings of connection and disconnection with past histories of trauma and violence. When examining testimonies of injustices and traumas, educators must explore the field of shame, not because knowledge of it "installs the proper guilt" (Britzman, 1998, p. 112). Rather, examination of the politics of shame in education enriches the possibilities for affective connections that engage someone in modulations of otherness.

The second challenge is to take into account that witnessing shame in the classroom is *collective,* that is, a collectivized engagement in learning to see, feel, and act differently (Boler & Zembylas, 2003; Zembylas, 2006b). It is precisely this collective witnessing that leads to a political awareness and attends to the work of transforming the future. It is in this way that the classroom can become a potential place of political transformation, when students are acknowledged as witnesses who work through their affective connections to the victims, themselves, and their cowitnesses and find their grounding in solidarity and conviviality.

Expanding on these small openings, I want to make visible the *therapeutic* aspects of shame, instead of getting caught in its *toxic* perspective. In particular, I have shown how a productive politics of shame in the context of intercultural education generates possibilities for questioning the process of national or racial identity formation that is construed through the distinctions between legitimate self and illegitimate Other. Thus, a politics of solidarity and conviviality can be made possible because the formation of the affective economies of exclusion/inclusion is not fixed on pride; rather, forms of shame are made available for critical self-reflection. Furthermore, in order to engage reflectively with the notion of shame in the way that it is suggested here, both teachers and students need to examine and understand the social and/or institutional structures through which subordination of the marginalized takes place. This would also provide an opportunity to understand the emotions underpinning these oppressive structures and in turn provide those engaged with the experience of reflection on shame with an opportunity to consider how structural and attitudinal change might be made. As it has been emphasized, shame and pride are deep rooted and complex emotions and challenges to people through these particular sets of feelings can be deeply disconcerting and can result in entrenchment rather than reflective growth.

Against a perspective of shame that would see it as something paralyzing, negative, and destructive, I have drawn out a different account of the politics of shame in intercultural education. I have shown that witnessing shame deepens the politics of identity, in spite of the resulting discomfort and ambivalence. Shame is neither *good* nor *bad*, or inherently transgressive (Probyn, 2005). Attention to shame, as Probyn suggests, allows us to reach beyond the normative categorizations of identity and otherness. In political terms, this idea means that shame allows educators to enact solidarity and conviviality among individuals and groups in ways that pride has missed.

Furthermore, this opportunity provides educators with the critical tools to move beyond the usual dichotomies in multicultural approaches between privileging either unity or difference. However, it is important to acknowledge that these critical tools are insufficient to the reframing of school life. The gap that exists in teachers' awareness and capacities to handle these dichotomies and what this might point to in terms of teacher development suggests that it would take a much more concerted approach for a productive politics of shame in intercultural education—one that places the affective more at the heart of teacher understanding and, ultimately, teacher education.

All in all, avoiding the unpleasant feelings of recognizing shame may well amount to avoiding the openings that are created in intercultural education to transcend the rigid, unitary, and static Other and establish powerful affective connections. The normalizing and totalizing tendencies of polarities around shame/pride easily stigmatize others as enemies, barbarians, or simply different; similarly, if unity is privileged, then the uniqueness of cultural communities is undermined. Utilizing the prospects of a productive politics of shame in intercultural education, however, keeps open the following possibilities: First, the others are not locked in the image we have constructed for them, and second, the others may show us something new and perplexing about ourselves that we have not recognized before. This kind of mutual or reciprocal engagement between various groups—that are constantly redefined in numerous terms—"avoids the kind of false consensus, reciprocity, and recognition" (Tarnopolsky, 2004, p. 485) that is often promoted in some multicultural discourses. A truly productive politics of shame in intercultural education is one that preserves the openness to this kind of ambivalence, which is so important to the experience of being shamed against any moralistic norms.

Chapter 6

Dilemmas of Justice in Peace/Coexistence Education (with Zvi Bekerman)

There has been considerable research conducted on ethnic (and other forms of) violence in the last 30 years, indicating how violence contributes not only to more fear, anger, and hatred in conflict-ridden communities at large, but also to poor student performance and troubled emotional health (e.g., posttraumatic stress) for students, teachers, and their communities (Harris, 2007).[1] As violence, trauma, and their emotional implications become issues that inevitably need to be dealt with in schools, teachers throughout the world confront enormous pedagogical challenges (Berlowitz, 1994). These challenges led some to turn their interest to peace studies and nonviolence education—a field that has grown considerably since the 1980s. Although such efforts take many different forms around the world, it is emphasized how educators who work in this area need to pay attention to the relationship among violence, trauma, and social inequalities in the world and thus consider the connections between peace and social justice (Burns & Aspeslagh, 1996; Harris, 2002a)[2].

In general, peace education programs explain the roots of violence, teach alternatives to violence, and empower students "to redress the circumstances that can lead to violent conflict" (Harris, 2004, p. 5). Peace education has many components and meanings within different settings (Danesh, 2006); yet one of its most important components is the promotion of a concept of peace accompanied by social justice (Harris, 2002a). As Burns and Aspeslagh (1996) argue, one of the features of peace education is its alignment with a radical/counter-hegemonic paradigm for social change through education.[3] Within the educational context, "a concern for social justice means looking critically at why and how our schools are unjust for some students" (Nieto, 2000, p. 183). Within a broader social context, social justice implies equity and access to opportunities for all, regardless of their differences; peace and coexistence means respecting social justice and working together to resolve conflicts (Harris & Morrison, 2003; Opotow, Gerson & Woodside, 2005)[4].

Lyotard (1988) argued that the major problem of our times may be understood in terms of two issues: the impossibility of avoiding conflicts and the absence of a universal genre of discourse to regulate them. "And so," as Lyotard and Thébaud (1985) write, "when the question of what justice consists in is raised, the answer is: 'It remains to be seen in each case,' and always in humor. But also in worry, because one is never certain that one has been just, or that one can ever be just" (p. 99). Lyotard sketches the idea that justice is based neither on universal criteria nor on particular content. His understanding of justice has no relevance to contractual justice; instead, he wants us to pay attention to the institutional practices and the social interactions that structure the experience of justice (Dhillon, 2000).

Drawing upon Lyotard's ideas, we face an important dilemma concerning the universality or particularity of justice—a dilemma that has consequences for peace and coexistence, as it will be discussed later in the chapter. On the one hand, if we accept that there are no universal criteria or norms of justice, then there is the concern of relativism and the lack of consistency; thus, it is argued that a common set or rules is important for establishing the law. On the other hand, if we insist that there are universal criteria of justice, then there is the danger of authoritarianism and the lack of attention to the particularities of moral experience. By reducing the complexity of experience to a set of guidelines mandating what is just in advance of experiencing the Other, the emphasis is placed on the epistemology of justice not on its ethics. A turn toward the ethical considerations of justice is a turning away from the metaphysics of justice and a turning toward the practice of justice.

We do not think that the choice is an either/or dichotomy between universality and particularity, but the ethical and epistemological considerations of justice can be located somewhere in between. Not "everything goes" but much more than universal monologism goes. While a set of parameters is important, we need to learn how to be attentive to context and history. Thus, justice may be placed somewhere between a responsive (justice as a response to the Other) and knowing (justice as exposure to (un)knowing the Other) ethics. The major advantage of this approach is that it does not reduce the complexity of human encounter and experience to an abstract set of rules. Rather, it points toward justice as the landscape of forming ethical practices that are conducive to coexistence with others.

Following this middle-ground approach has important pedagogical implications in relation to promoting peace and coexistence education (P/C-E), the focus of my interest with Zvi Bekerman in this chapter. Consequently, we consider that the choice between two major educational trends in extolling the importance of justice as a fundamental value to be transmitted to students—that is, *citizenship education* that highlights justice as a public value, and *character education* that champions justice as a personal trait (Kristjansson, 2004)—introduces another dichotomy that is problematic for us. Whether the focus is on citizenship or on character education, lurking in the background are deeper questions on issues around *identity* and *otherness*: How can justice be addressed in a world of uncertainty and multiplicity of identity that produces powerful emotions such as hatred, fear, and anger, as a result of defining one's identity through the negation of the Other? How can the dilemmas of justice create useful openings for the *ethical* and the *affective*—not only for the epistemological—implications of P/C-E? These questions are concerned not only with how to actively engage children to *know* how to be just in an uncertain world, but rather with how to *inhabit* affective spaces that seriously consider the fear and hatred for the Other in a manner that is vigilant to the emotionality of justice. The purpose of this chapter, then, is to turn our attention to the ways in which the emphasis on the pragmatics of justice is implicated in the politics of trauma in education and particularly to the affective politics of pedagogies that promote peace building and coexistence.

This chapter is divided into three parts. In the first part, we closely follow Lyotard's ideas to problematize claims about the universality of justice. Although we do not agree with everything that he says, we share his concerns about the pragmatics of justice and build on such ideas to ask further questions related to the emotional aspects of (un)just practices. Lyotard reminds us that justice claims are interrelated with identity issues and as such they create much emotional ambivalence and thus have significant affective implications. In the next part of the chapter, we examine the ambivalence of inhabiting various positions of justice through considering the politics of identity in the context of conflict and the efforts for peace building. Discussing the importance of *postpositivist realism* (Moya & Hames-Garcia, 2000), we emphasize the importance of appreciating the primacy of the interactional, affective, and material over the idealized and reified categories in shaping the ideologies on which legal, health, and political structures are founded and to which definite forms of social consciousness (identity/class/culture) correspond. Finally, we explore how ideas about the pragmatics of justice may enable educators to comprehend the politics of trauma in education and may encourage peace and coexistence practices.

NOTIONS OF JUSTICE AND AFFECTIVE IMPLICATIONS

Against modernist notions of justice, Lyotard focuses on the impossibility of a contractual consensus on justice since there are so many competing worldviews, and particularly, the lack of equal participation to the contractual process by those who are marginalized (Dhillon, 2000). Lyotard asks us to attend to claims about the universality of justice, and how they have been used to perpetuate oppression against the marginalized. Lyotard's views lead neither to relativism nor to passivity; instead, he urges us to engage in reflective judgment and respond to the responsibility we have for the Other's suffering. In responding to such responsibility—which takes precedence over thinking or knowing about the Other—one learns from others to encounter ethics as action. Only when response is accompanied by action, it acquires its full meaning. It is thus important that this word does not remain in the abstract or is disembodied, but that it emphasizes its material direction.

The ethical singularity of the Other makes us approach his/her suffering with humility and vigilance, emphasizes Lyotard. Thus, for Lyotard, justice is not the blind application of rigid criteria but the exercise of prudence (Lyotard & Thébaud, 1985); "it is proper to prescription to be left hanging in midair" (ibid., p. 26). This notion implies the *pragmatics* of justice, as Lyotard refers to it; that is, the idea that judgment occurs in practice and therefore one must constantly bear witness to difference and to the obligations we have to others. This pragmatics forces us to continually hold open the question of justice; in other words, justice is not an abstract ideal for Lyotard, but a continual agonistics to listen to the Other, respond to him/her, and constantly question one's self whether a decision/action is just. As he emphasizes:

> Absolute injustice would occur if the pragmatics of obligation, that is, the possibility of continuing to play the game of the just, were excluded. That is what is unjust. Not the opposite of the just, but that which prohibits that the question of the just and the unjust be, and remain, raised. Thus, obviously, all terror, annihilation, massacre, etc. or their threat, are, by definition, unjust. (Lyotard & Thébaud, 1985, pp. 66–67)

The pragmatics of justice is guided by the notion of the obligation to others and not the imposition of laws as though their contents were transparent (Todd, 2007). In other words, justice is established on the basis of the interpersonal encounter with the Other; thus, justice is the response to the Other. "And that response," explains Todd, "can only live up to its name of response when I refuse to impose upon them [the other individuals] a set of criteria or to hear their words only through a filter of laws, regulations, and such" (ibid., p. 597).

However, even when pragmatics is acknowledged, our worries are not over. Here we want to warn and remind the reader that taking action toward justice is no easy task. Maybe the two axioms associated with the hypocritical tradition can be of help regarding the when and how of action. *Curate ipsum* (Latin for "take care of your self") is the first that for our purpose indicates the need to deal with our own injustices before dealing with those done by others to others; and second, *primum non nocere* (Latin for "first do no harm") that might warn us that our activity directed to the correction of injustices should be first and foremost careful to do no other (new) harm.

In general, it is important to acknowledge that Lyotard's moves are made within a philosophy of language and not a philosophy of consciousness. Within his pragmatics of discourse, axiomatic laws are unable to provide any advance criteria for judgment of any language game (e.g., justice). Rules apply to games after the fact, that is, after games that have been played. Thus, in seeking to determine what a *just* decision is in the Israel-Palestine conflict or in the conflict in Iraq, one cannot follow a predetermined list of laws. Lyotard emphasizes that "any attempt to state the law, for example, to place oneself in the position of the enunciator of the universal prescription is obviously infatuation itself and absolute injustice, in point of fact" (Lyotard & Thébaud, 1985, p. 99). This idea does not imply that there is a plurality of justices, but rather a justice of plurality, as Lyotard asserts: "Can there be a plurality of justices? Or is the idea of justice the idea of plurality? That is not the same question. I truly believe that the question we face now is that of plurality, the idea of a justice that would at the same time be that of a plurality, and it would be a plurality of language games" (Lyotard & Thébaud, 1985, p. 95).

The problem in Lyotard's work of reconciling the "plurality of justices" with the "justice of plurality" is a formidable one (McLaren, 1994). "To judge without criteria, as Lyotard urges," writes McLaren, "affirms the utopian imagination as the grounds for making ethical decisions" (p. 333). Although McLaren is right, one cannot help but notice that the perception of having to decide between a "plurality of justices" and a "justice of plurality" suggests a dichotomous epistemology. It is worth remembering that the either/or options are strongly related to Western philosophical developments that stand at the basis of what we are discussing here. What we need theoretically as well as practically is to find ways to abandon these dichotomous perspectives.

Clearly, the type of justice advocated by Lyotard is the "justice of heteronomy, of irreducible difference. This justice is *not lawless*, but it does not legislate. It results in an 'unresolvable dissensus' in which no individual is subjected to a law that is alien to him or to her" (McLaren, 1994, p. 330, emphasis added). Therefore, it is obligation, imagination, and flexibility that watch out "for prescriptions that appear just or authorized" (Lyotard & Thébaud, 1985, p. 66). By endorsing a justice of plurality and difference, Lyotard reinforces multiple identities while refusing to offer a standpoint epistemology from which universal positions may be adjudicated (Best & Kellner, 1991).

Undoubtedly, there is something deeply unsettling in Lyotard's attempt to advocate a respect for difference and obligation to others as an antidote to the normalizing conventions of contractual justice and the rigidity of self-identity attached to particular conceptions about the nation, religion, or race. The turn that he takes is not informed by rationality or epistemology alone. Rather, it is a practice that promotes recognition for the obligation to respond to the Other, *relationally* and *affectively*, and this is something we want to retain in our position as well. There may be no foundational criteria of justice, but this need not rule out provisional agreements for coexistence and the construction of warranted assertions about what constitutes suffering and trauma—that is, the recognition that justice as a response to the Other unavoidably involves the navigation in difficult affective spaces.

In other words, how affective relations bound up with stories of (in)justice has important implications and educators need to be able to construct spaces in which difference and identity can be understood and lived relationally and affectively. As Ahmed (2004) and Berlant (2000) point out (see chapter 1), there is a difficult relation between emotion and (in)justice. On the one hand, injustice cannot simply be reduced to feeling bad (Berlant, 2000); on the other hand, justice is certainly related to feelings but it is not about having the right kind of feeling (Ahmed, 2004). To make a connection to Lyotard's ideas, justice is a sign of obligation to others, through the reification of feeling for others and their suffering. This is why injustice may take the form of perpetuating particular kinds of affective relations to social norms (Ahmed, 2004). Challenging injustice also implies the interruption of these norms and consequently the subversion of the affective relations to these norms. Lyotard reminds us that justice and identity claims must be constantly interrogated and continually reinvented outside of universal statements that are epistemologically driven. To attend to difference in education is not to "speak of it" but to offer possibilities for students to experience the impossibility of responding to others within a set of absolute criteria; in this manner, students will experience the ambivalence of inhabiting various positions of justice.

CONFLICT, PEACE BUILDING, AND THE POLITICS OF IDENTITY

One way of examining the ambivalence of inhabiting various positions of justice is to consider the politics of identity. Given that self-identification is inextricably intertwined with a large-group identity, the study of conflict is a key link between individual and large-group psychology, because the politics of trauma reveals intergroup suffering. It has already been shown in previous chapters that the politics of trauma is perpetuated because each group continues, through various means, to delegitimize the actions and beliefs of the other. Waller's (2002) study of communal conflict illustrates the polarizing nature of narratives about the Other:

> Instead of judging other people across a broad spectrum running from good to bad, we make extreme categorical judgments based on the polar opposites of "good us" versus "bad them." Our case is sacred; theirs is evil. We are righteous; they are wicked. We are innocent; they are guilty. We are the victims; they are the victimizers. It is rarely *our* enemy or *my* enemy, but *the* enemy—a usage of definitive article that hints of something fixed and immutable, abstract and evil. (p. 243)

What needs to be considered is that this us/them monological divide is not something *natural*. For an us/them divide to make its appearance in the world, historical processes developed that, though not only identifiable in the Western world, are tightly connected to the processes of modernization. These processes have been established on two major ideas: first, the centralization of power and second, its development both into colonialism and the solidification of the nation-state (Bhabha, 1990; Elias, 1982; Foucault, 2000; Gellner, 1997). For identities and cultures to become available, they need to be functional in given contexts; as such, they have come into being in societies that define and organize them. Identity/culture helps individuals and groups to be situated.

Undertaking a long critique of the concept of identity in its traditional positivist psychological meaning is unnecessary, as a great deal of academic work has been devoted to this issue (see Bakhurst, 1995; Gee, 1992; Hall, 1996; Harré & Gillett, 1994; Potter & Wetherell, 1990; Watt, 1997). These studies point to the relatively modern appearance of this concept and its close connection to sociohistorical and philosophical developments in the last 400 years of Western intellectual history. Identity as a unitary and autonomous construct has come under attack as being a product of exclusionary power relations (Bhabha, 1994), a monologic posture that tries to overcome through domination that which is *by nature* dialogic: the self and identity (Bakhtin, 1981; Gee, 1992; Mead, 1934; Sampson, 1993). Similarly, social identity and its constitution have been analyzed as the product of power relations that establish dichotomous hierarchies (Laclau, 1990) in which the powerful attain the status of essentiality while the weak are reduced to the rank of an unfortunate but necessary accident (e.g., man/woman, black/white, Jew/Arab).

The powerful machinery developed by the nation-state mostly in the shape of massive educational efforts that market universal (anonymous) literacy has been successful in making seem natural or banal, as Billig (1995) would have it, the detailed practices through which nation-states become almost invisible settings. In these settings, we *mistakenly* hold a sense of individuality—an individuality always measured against a contingent Other (Laclau, 1990) and the modern court of human appeal: the *high* culture of the nation-state (Williams, 1961).

Theoreticians have identified the national structure as one of the cruelest systems on the historical scene (Bhabha, 1990; Mann, 2004). For the community to be imagined in its national oneness (Anderson, 1991; Hobsbawm, 1983), borders had to be widened and groups lumped together through homogenizing efforts; culture had to be reified and the individual, and one's relation to the sovereign, strengthened so as to undermine the power of smaller communal identifications. Concealed behind the promise of universal equality was the sovereign's demand to have no other than an individual, stripped of any group affiliation, under one's rod (Mendus, 1989).

It seems to us that educational approaches to conflict situations are lacking in that the reality of the nation-state is not sufficiently accounted for ideas that contribute to our understanding of identity in cross-cultural encounters. Billig (1995) fittingly points to the flattened topography of identity theorizing shaped by the fact that nationalism is overlooked by identity theoreticians and made, by default, functionally equivalent to other types of identity. Reicher and Hopkins (2001) urge the social sciences to recognize the crucial influence of national structures in the shaping of group identities. This oversight is detrimental to social theory in general and more so to that which is concerned with educational interventions.

The central question we ask then is: Can change be really achieved without first exposing and overcoming the structures and practices that have established the present conflictual situation and their functional categories? The risk, as Appiah would have it, is that we end up replacing "one kind of tyranny with another" (Appiah, 1994, p. 163). The call of a politics of identity is not much more than a mirror image of that same political process that erased identity by universalizing it. Just as dominant groups in the culture at large insisted that the marginalized integrate by assimilating to dominant norms, so within some practices of identity politics dominant subgroups may, in theory and practice, impose their vision of the group's identity onto all of the group's members. This should come as no surprise, when considering that education is a social process conformed by the same social forces we are trying to change. When confronting this problem, we immediately realize that, if at all, change is no easy task.

Consequently, strengthening peace and coexistence might not be achieved if alternative options to the ones dictated in the past are not pursued. Though we worry about the possibility of replacing one universal position with another, we want to be less pessimistic and not give up easily on "all of us"; still how to achieve this is not easy to realize. A stipulation is in place, then. The monological pattern of identity/culture construction is paradoxically dialogical and based on the negation of otherness. In this sense, the sane and the pathological have built their identities/culture also through a dialogue that negates each other. However, dialogue, one way or another, seems to stand at the basis of change.

In our efforts to respond to the question posed earlier, we have recently considered what has come to be called *postpositivist realism*

(Moya & Hames-Garcia, 2000). This view, though acknowledging poststructuralist critiques (such as Lyotard's views), realizes that goods and resources are still distributed according to identity labels, be they ethnic, national, religious, or what is regarded as pathological. Our approach asks for the realization that poststructural theoretical conceptualizations, though important, in that they seem to us to serve as a better description of the ways we experience and shape our present environments, might not influence the world to the extent that the constructed perspectives these theories presume. Poststructuralism might be theoretically *right,* providing a good description of an empirical world, but it is no stronger than the powerfully constructed *reality* of a consequential hegemony that significantly affects our lives and sets the limits for where we can live, whom we can marry, what educational and employment opportunities are available to us, and the location of the boundaries of that which is considered to be normal. Allowing for the recognition of what identity and culture *do* in real life might sooth the suspicions our arguments may raise.

Though this strategy should be functionally useful, we should be aware of the price we might pay when acknowledging its influence on our educational work. That is to say, we should not work toward instilling/strengthening in students modern positivist perspectives of problem identification through idealist, essentialist, and causal visions, subtly disguised or not, in our discursive practices in education. By this, we mean that we should try and not allow *reality* to confuse us as to what the problem is. The problem is not only the marginalized student but also the *normal* that marginalizes him/her. The great danger in acknowledging reality is, in our specific case, confusing it with that which points at the Other as its problem, thus usually failing to offer a solution because of remaining attached to a monological perspective. The question is clearly how to make us aware that we all partake in the constitution of the problem and thus become constitutive of its solution; that is to say, how to help us realize the dialogical nature of that which is presented as monological. From our perspective, a relevant path implies a radical change of the educators and the educational activity. The individual mind can no longer be the goal of these activities, as is the case with an education aware of cultural context, namely the minds of the other individuals that conform to these contexts (parents, teachers, etc.). Instead, we ought to consider and analyze the interactional practices and strategies through which all involved in the contextual activities allow the Other to make its appearance.

Simply stating this as a fact will not do. We first need some criteria that can be accepted by all and, in this all-sense, that are *objective* but do not rest on the positivist underpinnings of objectivism. Such criteria could be achieved through an educational process directed at questioning who in our societal context are exploited. Who are powerful and who are powerless? Which cultural patterns carry symbolic power? (Hames-Garcia, 2000) Who are labeled as sane or as ill, good, or bad? Who define normality or justice?

The move suggested is similar to that taken in anthropology (Varenne & McDermott, 1998) when examining central paradigmatic perspectives in general education—its study and practice. It starts by restoring the concept of identity/culture to its historical sources, thus de-essentializing it. It follows by developing the restored meaning into a methodology—cultural analysis—that is to say the gaining of skills on how to read/describe the world through careful observation and recording of practical activity, which in turn allows for a shift from the individual or the socializing group as the crucial analytic unit for (educational) analysis to the production of cultural contexts through social interaction. And last, it leads to a new articulation of major policy issues related no longer to identity/culture and its components (individual, texts, etc) but to the analysis of particular identities/cultures and how these are produced/constructed in the particular context of particular societies. Looking at the world in this way, seriously and critically, means being open to finding new criteria through which to name phenomena. The process could be liberating in that it could bring about the understanding, again, that identity/culture are not necessarily the right criteria through which to describe the world, its inhabitants, and events; not that they do not necessarily exist or are only hegemonic constructs but that though they are legitimate, they not necessarily need to imply individual suffering.

If successful, we might come to appreciate the primacy of the interactional, affective, and material, over the idealized and reified categories in shaping the ideologies on which legal, health, and political structures are founded and to which definite forms of social consciousness (identity/class/culture) correspond. This implies bringing the social and affective to predominate over the ideological and thus opening the potential of collaboration rather than individuation.

TEACHING ABOUT JUSTICE AND THE POLITICS OF TRAUMA IN THE CONTEXT OF P/C-E

Considering justice, peace building, and the politics of identity in the way discussed here may enable us to comprehend the politics of trauma in education and how its interrogation may encourage peace and coexistence practices. The usual focus on the epistemological aspects of P/C-E—that is, the emphasis on what "to know" about the Other—has tended to overlook the interactional, affective, and material aspects of justice and peace building. But exploring the affective politics of P/C-E begins to create a connection between individual psychic patterns and the broadest accounts of social change. After all, the idea that emotions and affects—as actions and practices—are the very site of the capacity to effect change should not be underestimated (Lutz & Abu-Lughod, 1990). In the context of P/C-E, to take seriously the relations and powers of affects is to make such practices part of the effort to articulate a *critical politics of affect* (Zembylas, 2007) that involves the explicit mobilization of affect; this mobilization aims at producing new ways of inhabiting spaces of social encounters that promote *just* practices.

Hesford (2004) wonders what forms of engagement are constituted as affective responses to injustice, violence, and trauma and what the limits of such witnessing are. Therefore, she uses the term *crisis of witnessing* to refer to "the risks of representation" of injustice, trauma, and violence and "the possibility of empathetic meaning between witness and testifier, listener and speaker" (p. 107). A critical approach to the crisis of witnessing urges us to question the assumptions that completely erase any distinction between the self and the Other. After all, witnessing (in the simple sense of just seeing a situation) does not necessarily entail an empathetic connection, and thus any understanding of affect should take into account that people will have different affective responses, some of which do reinforce the self/Other dichotomy (Zembylas 2006b). But more seriously, perhaps, the presupposition that one party involved in witnessing can *feel* the Other not only sentimentalizes witnessing through a misleading belief in ethical and political symmetry, but also risks incorporating the Other within the self. Undoubtedly, empathy is never enough and if alone might at times be insulting and patronizing—not to mention that empathy by itself may entrap us to the axiom *primum non nocere*. On the other hand, empathy as an activity in the world (not in the mind) should not be completely ruled out as a possibility of responding ethically to others. This kind of empathy includes a critical effort to accept the Other as a distinct individual and thus challenges the process of dehumanizing him or her (Halpern, 2001).

When adopting such perspectives, educators should make sure they do not adopt them so as to acknowledge *truth* and *objectivity* to be always socially and discursively constructed, mediated by power discourse and desire, and thus in a sense totally relative. Doing this does not imply giving up on change or, worse, justifying present unpleasant situations. The question of suffering and trauma is not theoretical—at least not for those who suffer. Though the foundational account of objectivity should rightfully be critiqued, we want to suggest that it is necessary to sustain what is epistemically and affectively valuable in the concept (Mohanty, 1997, 2000). From a pragmatic postpositivist perspective, objectivity is to be understood to be essentially theory-dependent and socially situated. Postpositivist realism acknowledges the mediated nature of language and reference, but it also argues that we can still come to know something about the world and feel its injustices, to distinguish truth from error, fact from fiction, and even the normal from the pathological. From this perspective, objective knowledge is gained not by rejecting all forms of bias (which is the positivist take) but rather by examining and taking into account the epistemic and affective consequences of bias (Moya & Hames-Garcia, 2000; Nguyen, 2000). This is central to an education that wants to consider the dilemmas of justice within a pragmatic, postpositivist perspective.

Thus, we should be aware that when perspectives become totally relativistic they are rather irrelevant in educational work (Bekerman, 2007b). While acknowledging positivist limitations, a pragmatic postpositivist approach attempts to effect change in the world. The outcome of such an approach should work by defining the new uncovered perspectives as challenging the present world through the shaping of new accounts of social relationships and affective connections (asking for example: What types of transformations do we need in our values and behavior? How can social institutions, such as schools, make a real difference in the lives of the poor or underprivileged?) (Bekerman & Tatar, 2005). Bearing witness to the politics of identity—for example, through examining narratives of suffering within a community and its claims that justice needs to be served—signifies the demand for new affective connections. It is precisely in the possibility of such connections that teachers and students are called into being witnesses of testimonies. A critical engagement with testimonial narratives means that teachers and students have to decide how to become critical witnesses of these testimonies (rather than merely spectators, consumers, or tourists) and consider how these testimonies change (if they do) the feelings about the (marginalized) others. To push this one step further, we should also consider the possibilities for educational activities that try to critically approach the concepts we have constructed in this world and ask if they are at all useful. What we are arguing for at this point is that what might be needed is not for the victim to be understood, recognized, and included (or not only this) or for the victimizer to become more sensitive and understanding, but for both to *understand* and *feel* the world differently.

Lacking an ethics and politics of affect in P/C-E "flies in the face of all the evidence that human individuals [. . .] only exist as faint traces in much larger and more extensive circuits of social relation" (Thrift, 2004, p. 67). Peace/coexistence education has to be understood as producing not only epistemological but also ethical and political subjects that are located within certain social institutions and specific political contexts. In linking affect to such contexts, pedagogy can begin to problematize the ideological subordination of individuals and groups and how the affective organization in school, family, and the workplace best supports the existing social order.

Any viable ethics and politics of affect in P/C-E needs to be sufficiently flexible to acknowledge the sociopolitical configurations of power relations, underwrite the complexities of critical pedagogies, and recognize the unavoidable ambivalent character of affects as both revealing and concealing (in)justices (Zembylas 2006b, 2007). An ethics and politics of affect in P/C-E reveals and conceals the world in many ways. The challenge for educators in creating opportunities for meeting and responding to such (in)justices is to help students experience affect as a creative source of personal and political insight. This idea does not imply *working on* students' heads—as if the preceding challenge would be solved by changing students' cognitive content—but rather *working with* students to change their (and our) worlds through just relations. Critical witnessing, therefore, challenges us to question the construction of ethical and political relations and imaginaries through testimonial practices.

Coming to understand and practice justice based on a pragmatic, postpositivist perspective in the context of peace and reconciliation education, dialogue is one way teachers and students can find out which emotions they are experiencing as "different interpretations produce different awarenesses and understandings of our emotions" (Burrow, 2005, p. 35). Here the challenge to teachers and students is to begin to question their own affective investments and practices and to begin to see how these investments/practices mask various forms of (in)justice. Thus, the following questions may be asked: Do these affective investments/practices permit teachers and students to enter into dialogue? Can teachers and students form a classroom community that supports the expression of each other's experiences through becoming empathetic listeners? Do such affective investments/practices produce ethical response and sharing and generate bonds through strengthening the space of relations? The possibility of a relational affectivity within what Burrow (2005) calls *communities of interpretation* points to the subversive disruption of fear, hatred, and resentment. The challenge is how an educational process of this sort works practically toward political action; this could be achieved through plenty of actions such as demonstrations, petitions, publications, and so on. An example of how an educational process of this sort works practically toward political action is described in the next chapter.

Chapter 7

Mobilizing Anger for Social Justice in Education

I speak out of *direct and particular anger* at an academic conference, and a white woman says, "Tell me how you feel but don't say it too harshly or I cannot hear you." But is it my manner that keeps her from hearing, or the threat of a message that her life may change? [. . .] Women responding to racism means women responding to anger; the anger of exclusion, of unquestioned privilege, of racial distortions, of silence, ill-use, stereotyping, defensiveness, misnaming, betrayal, and co-optation. . . . Every woman has a *well-stocked arsenal of anger* potentially useful against those oppressions, personal and institutional, which brought that anger into being. Focused with precision it can become *a powerful source of energy serving progress and change.* (Lorde, 1984, p. 127, emphasis added)

Anger is often described as a *dangerous* emotion that threatens rationality, social order, and constructive dialogue (Jaggar, 1989; Lyman, 2004). Therefore, there are several approaches in many aspects of social and professional life emphasizing *anger management* based on feeling rules that are socially and culturally determined (Goleman, 1995; Hochschild, 1983). In particular, the goal of anger management—which has become a widespread movement in schools, organizations, and the workplace in the West and especially in the United States (Morris & Feldman, 1997)—is to control one's emotions, that is, repress, neutralize, or even express anger occasionally but do it in *appropriate* ways (Tracy, 2000).

The definitions of *anger* vary, of course. However, my concern in this chapter is with anger as a response to a perceived injustice; what has become known as Aristotle's *moral anger* (Boler, 1999; Stocker,

1996). As Lyman (1981) writes, "one can define anger as the essential political emotion" (p. 61), because it motivates people to raise their voices against injustice and can be used to inspire transformation and social change (Lorde, 1984; Spelman, 1989; Swaine, 1996). When anger is taken seriously as a communication, rather than being identified with an inner, psychological disorder or inappropriate and uncivil behavior, then becoming witnesses to those who feel they have lost their right to have a voice has the potential to enrich dialogue and enhance the ability of educational institutions to address injustice.

Undoubtedly, anger (and violence) can be highly destructive inside and outside educational institutions. However, here I want to problematize the fact that the significance of moral anger in education is essentially eradicated, when a distinction with other forms of anger is not made. Unfortunately, this is usually the case, because more often than not all anger is deemed as socially inappropriate as soon as it is expressed. Admittedly, how to make a distinction between moral anger and other kinds of anger (e.g., road rage, vengeful anger) is not a simple task. However, an important aspect that distinguishes moral anger from other kinds of anger is the notion of someone becoming angry as a witness of gross violations of justice, humanity, and dignity; in other words, moral anger is what motivates someone to oppose injustice.

But what are the dangers and possibilities of encouraging or prohibiting the creation of educational spaces in which moral anger can be overtly explored (i.e., anger as a mode of analysis) and expressed? Does this inevitably contribute to further polarization, or can it challenge injustice and allow for greater empathy and understanding? On the one hand, silencing anger in the classroom because of its supposed *irrationality* is ideological; it justifies and perpetuates domination "by silencing the voices of the oppressed, labeling anger as 'loss of control,' as 'emotionalism,' or as 'neurotic' " (Lyman, 2004, p. 134). On the other hand, as Holmes (2004a) suggests, anger is not inevitably emancipatory but ambivalent, because it is part of an ongoing struggle of power relations. Thus, encouraging explorations of anger at injustices and inequalities can inspire students and teachers "towards new considerations of their relations with others and open up new spaces for continuing struggle" (p. 223).

In this chapter, I use feminist and poststructuralist discourses to suggest that the politicization of anger in education is not only inevitable but also desirable. In particular, an analysis of anger in education may offer certain critical advantages, including a better understanding of the importance of *being angry* in political terms. As various educational approaches nowadays seek indiscriminately to silence anger in the name of "emotional intelligence" (and thus, depoliticizing it), an attention to the political aspects of emotions in education is extremely valuable. This chapter, then, seeks to establish some of the linkages connecting the politics of trauma, injustice, and anger; it highlights the philosophical issues that arise when one considers a politics that imagines anger as an alternative form of relationality. This form of relationality is not stuck in either victim politics or revenge, but involves a process of working through of traumatic memories and experiences that open up the possibility of overcoming the cycles of victimization or revenge that are usually acted out in various forms of violence.

Therefore, I will argue that anger is central to the exercise of power relations in the classroom; for this reason, I will challenge the dominance of instrumental/rational accounts about anger replacing them with the recognition of anger as political. Also, I will clarify some of the conditions that generate anger as ambivalence—clearly, I am not arguing that we are to see *all* anger as ambivalent—and explore the transformative possibilities that this ambivalence creates (see also chapter 3). I believe that this work is important, because there has been little attention to anger in relation to educational politics. Much of the scholarly attention has been given to recognizing anger as one of the emotions that teachers and students experience in classrooms; however, the significance of the emotion of anger in social justice education has not been adequately analyzed.

The first part of this chapter explores the dominant discourses of anger in Western societies and the implications for education. The second section sketches the politicization of anger in education and its importance for educational purposes related to social justice. Finally, I explore how an educational politics of anger might be possible, by seeking to create spaces in which students and teachers can develop empathetic dialogues and relish the transformative capacity of anger. An educational politics of anger does not intentionally seek to cause anger, but to encourage students and teachers to develop a critical analysis of anger, as it is likely that they will experience such feelings when they begin questioning long-held assumptions and beliefs about their own traumatic experiences, other people, and social/political events.

Discourses of Anger in Western Societies: The Implications for Education

Holmes (2004b) argues that there are three major discourses of anger in Western societies: (a) The expression of anger is good in itself, but it has to be highly managed; (b) there are dangers in repressing anger; however, the direction of anger has to be harmless; and (c) it is good to repress anger; it is an indication of civilized individuals. The common characteristic of all three discourses is the assumption that there is supposedly a dichotomy between reason and emotion; anger denotes a lack of rational control and thus it eventually needs to be managed. In this section, I show how these discourses are also present in contemporary educational ideals and discuss their implications.

The first discourse has its roots in Aristotle's *Nicomachean Ethics* and the *Rhetoric* (Stocker, 1996) and can be traced through to contemporary times in references about *emotional management* and *emotional control.* Aristotle offers a cognitivist theory of emotion, that is, a theory in which emotions are linked with beliefs and reasons (Boler, 1999). Thus, anger, for example, is an emotion *about* something and denotes an evaluation of something. In the *Nicomachean Ethics,* a good person (*phronimos*) must be capable of getting "angry at the right things and with the right people, and, further, as he ought, when he ought, and as long as he ought." Anyone who does not get angry when there is reason to be, according to Aristotle, is a dolt. In the *Rhetoric,* Aristotle defines anger (*orgē*) as "a desire accompanied by pain, for a conspicuous revenge for a conspicuous slight at the hands of men who have no call to slight oneself or one's friends."

Three important points—that are still held centuries after Aristotle—can be made here. First, anger is understood in *moral* terms; it is aroused because men's dignity, respect, and honor are injured. Thus, the distinction between *moral* and *defensive* anger, for example, can be made. As Boler (1999) explains, defensive anger is what I feel when I perceive to be threatened by something, whereas moral anger is what I feel when I perceive something as injustice. In a previous work, Megan Boler and I (Zembylas & Boler, 2002) quoted a student criticizing peaceniks who "hate America so much" and saying that "America has done more than any other nation in the twentieth century to ensure the spread of freedom and democracy around the world." This student was described as exhibiting defensive anger, which "should be interpreted not so much as a righteous objection to one's honor, but more as a defense of one's investments in the values of the dominant culture." Defensive anger can be described as a protection of one's beliefs, values, and sense of identity; thus, fear of losing those is a fundamental feature of defensive anger. This interpretation begs the whole question of anger that is turned on the self and the expression of anger.

The second point has to do with the issue that anger concerns men not women. Feminists have shown how this kind of understanding anger has been used to legitimate men's anger but to slam and control women's anger to maintain their subordination (Bartky, 1990; Campbell, 1994, 1997; Spelman, 1989). This issue raises further questions about the genesis of anger and why some triggers for anger are seen as more acceptable causes than others. The question of who gets to decide what counts as acceptable anger—especially as this decision relates to gender—and how educators can cultivate this inquiry poses a real challenge. Finally, the third point refers to claims that dignity, respect, and honor denote the importance of social relations. This notion assumes anger as relational, something that people do (or do not do) as part of interacting with others; this is precisely what shows the importance of anger in the politics of everyday life (Holmes, 2004b).

In contemporary times, it has been Hochschild (1983) who first provided a comprehensive account about the need that anger be managed based on feeling rules that are historically and culturally specific. Also, in their classic study of anger in American life, Stearns and Stearns (1986) describe that anger is discouraged in contemporary America, unless it is highly managed. This view about anger management privileges rational, cognitive control of emotions, something that is also emphasized in Goleman's (1995) international bestseller *Emotional Intelligence*. Unfortunately, many calls for developing emotional intelligence and emotional literacy in schools aim at adapting students to the imperatives of *performativity*—how students should perform emotions (especially anger) in *appropriate* and *normal* ways (Zembylas, 2005). This is shown to be in line with business and management discourses to package "marketable solutions for success and self improvement" (Boler, 1999, p. 65). In our previous work, Megan Boler and I have shown that many contemporary discourses about emotional literacy in schools seem to get caught in the obsession for performativity, efficiency, cultural assimilation, moral self-control, and normalization of emotional skills. Consequently, anger management is largely framed as a set of competencies, precisely in the spirit of Goleman's work. Therefore, it is unlikely that anger management programs or emotional education curricula more generally make any clear distinction among various kinds of anger and particularly moral anger.

The second discourse about anger is related to psychoanalytic (especially Freudian and neo-Freudian) writings referring to the dangers of repressing anger. The similarity with the previous discourse is that this one also promotes the expression of anger, but it does so through an encouragement of *safe* expressions (Craib, 1994). Anger is depoliticized, that is, it is stripped from its social and political context and draws its legitimacy from the *psy* disciplines (Rose, 1998). Thus, pathologizing anger allows for the flow of regulatory programs and techniques (e.g., see Davies, 2000) that aim at domesticating anger by redirecting its energy to socially *appropriate* goals (Lyman, 2004). The issue again is, *who* gets to decide what counts as appropriate anger? This is where feminist and poststructuralist ideas will be particularly helpful later in this chapter. The language of repression and psychoanalysis may be useful, but it is not apolitical; as Rose (1998) has argued, utilizing the scientific authority of the psy disciplines aims at legitimizing the discipline of the subordinates' anger.

In the field of education, *moral education*—as it is exemplified in various emotional literacy curricula, particularly in the United States—is precisely an attempt to mobilize anger only to justify a new conception of the moral person (Boler, 1999). The right to anger is a resource for the dominant, a mark of power (Thorne & Henley, in Lyman, 2004), while angry speech by the subordinate is considered impolite and rude (see also Boler, 2004). Students, for example, have to learn how to (re)direct their anger using safe *techniques* that do not harm or offend others. Managing anger using various techniques is included among the topics in various emotional literacy curricula (see Goleman, 1995, 1998; Rae, 2003; Sharp, 2001). Anger management includes monitoring self-talk to catch negative messages such as internal put-downs, and realizing what is behind a feeling (e.g., the hurt that underlies anger). Anger management techniques possess a *gaze* that regards the student as an object and suspends any social or reciprocal relationships "in the name of disinterested rules of method" (Lyman, 2004, p. 138).

Finally, the third discourse about anger has its roots in the stoic philosophical tradition in which a major belief was that emotions needed to be wisely controlled. Stoics emphasized that the goal of life was to achieve tranquility of mind. Therefore, one becomes able not to feel anger because he or she is not subject to passions and reaches apathy (*apatheia*), which is both a mark of wisdom and a divine attribute. Again, this interpretation is not apolitical, because as Spelman

(1989) argues, it is oppressed groups who have been historically encouraged to repress their anger. After all, it is the dominant that requires the silence of the oppressed. Also, as Lyman (2004) notes, repression of anger implies acceptance of the dominant authority, "by denying the existence of an inquiry, or by lowering expectations of equal treatment, or by a loss of willingness or even ability to speak" (p. 138).

Women, for example, have been historically told that they are unable to express their anger overtly. This has been justified on a number of reasons, including that it is *good* for them not to express their anger (Campbell, 1997). However, there is a relationship between the expression of anger and the material context in which the expression takes place; therefore, the issue of politics and power relations is again central and makes one wonder who decides what is appropriate to express and what is not. In essence, what ideologies govern the *interpretation* and *expression* of anger in particular historical and social contexts? More relevant to education is the question: How have such ideologies of repressing anger as something inherently *good* been created, disseminated, and accepted in contemporary (emotional literacy) curricula?

Notice that the above three discourses are not contradictory and do not exclude one another. Repressing anger and expressing anger *safely* or *wisely* leaves precisely the necessary vague ground for normalizing young souls into managing their anger appropriately. There is certainly no distinction made between defensive anger and moral anger. Anger may be seen as useful, under some circumstances, but is dangerous; and as so, it has to be controlled. Also, in the above three discourses, an analysis of anger is embedded in particular power relations and therefore, the political implications of anger are unavoidable. Definitions of anger, expressions of anger, and interpretations of anger are situated in specific historical, ideological, and social contexts (Grasso, 2002).

My argument is that the political implications are not only unavoidable but also desirable to a degree. They are desirable, if educators wish to inspire vigilance to their students so that both teachers and students respond to unjust violations of the self and the community in which teachers and students live. As Grasso (2002) writes: "Because anger can be regarded as 'an emotion of defiance,' it challenges the attacker's power by asserting an equal claim to desire; and because anger can be regarded as an emotion that forces confrontation, it has the potential to disrupt existing power relations" (p. 10).

During my teaching career, I had the opportunity to experience several instances of this latter phenomenon in which anger provided the potential for challenging social events that seemed to take for granted the inequality of power relations. My students expressed how their anger about injustices in their lives had brought them to a cross-road in which they had to take some action. In one such occasion that happened in an American university a few years ago, after a heated discussion about the war on Iraq, an Iraqi student's anger managed to mobilize a small demonstration in the community against the war. An American student who participated in this demonstration explained afterward: "I gradually became so angry at the events taking place in Iraq that I had to cry out against what I considered wrong." This student continued by saying, "For a long time I felt guilty because I wasn't doing anything. After I realized Hasan's [the Iraqi student] anger and his perspective on the events, I knew that I had to do something." The students' expression of anger gave birth to a small but considerable movement to raise a voice against what they had considered as injustice, foster alliances, and form a radical reconception of self and agency—as is evident from this student's words. This example focuses on the expressed anger, although much anger is unexpressed yet still has social consequences. Nonetheless, I share Lorde's (1984) conviction that anger can be a positive and powerful source of personal and political insight in education. This is where I focus my analysis in the following section.

THE POLITICIZATION OF ANGER IN EDUCATION: AMBIVALENCE AND SUBVERSION

Hochschild (1975) has used the term *feeling rules* to refer to social norms that "define what we should feel in various circumstances" (p. 289). Such rules differ from context to context, indicating what is acceptable and what is not. We know these rules, Hochschild argues, from how others or we respond to inferences of emotional display. These rules, just like other rules, delineate a zone within which certain emotions are permitted and others are not, and can be obeyed or broken, at varying costs. For example, emotions of anger are often demonized and rejected as pathological and therefore, the feeling rule is that they should be suppressed or repressed.

Holmes (2004a), however, argues that the politicization of anger needs to go beyond a simple exploration of feeling rules "by addressing the *ambivalence* of anger and how it might make, rather than simply emerge from, subjects and the relations between them" (p. 211, emphasis added). Holmes explains that the notion of feeling rules does not capture the unruliness of emotions and the ambivalence of anger. In light of this claim, I want to raise some relevant questions: How can anger signify ambivalence? Are we to see *all* anger as ambivalent? How does ambivalence allow anger to be transformed into something constructive in an educational context? In this and the following section, I make an attempt to sketch a response to these questions.

Building on the discussion I have started in chapter 3, ambivalence refers to the coexistence of two opposite attitudes or feelings (conscious or unconscious), positive and negative (e.g., anger and joy), toward someone or something. Anger, for example, may be ambivalent because it is not only *destructive* (which is how anger is mostly perceived) but also *constructive,* as it can be directed toward achieving social justice, a change for the better. However, it should be clarified that not all anger is necessarily ambivalent. Whether anger is ambivalent or not depends on the particular circumstances that arouse one's feelings of anger. For example, Palestinian anger about the oppression by Israel may stem from a righteous indignation and thus one may argue that such indignation is not ambivalent. On the other hand, one should not exclude the possibility that these feelings of anger may also acknowledge feelings of empathy and understanding for the families of the victims who are killed in a bombing attack by Palestinian suicide bombers. The question is how one takes advantage of the opening that is offered in the ambivalence of anger as a means of raising a voice against injustice, yet in a way that is not stuck in victims' politics or revenge.

Therefore, if the ambivalence of anger is acknowledged, anger can be used as a positive mode of relationality. In other words, there seems to be a potential in recognizing and integrating rather than dichotomizing positive and negative modes of anger. Such integration of positive and negative modes of anger may be seen as a creative source of change, because it creates possibilities for constructive relationships among individuals and groups through which the complexities of differences are recognized and valued. Inevitably, then, the ambivalence of anger must be analyzed within particular social and cultural contexts. This ambivalence means that destructive anger is a performance requiring a reiteration of norms and the continued exclusion of that which exceeds certain definitions of anger. But because anger involves making judgments about what constitutes injustice, what expressions are appropriate, and under what circumstances, it is a *political emotion* (Lyman, 1981) and thus it has the potential of subverting normalization. That is, to reiterate an idea analyzed in previous chapters, anger is not a psychological property of individuals any more, but an indispensable aspect of a communicative negotiation that produces subjects and relations (Foucault, 1983).

The ambivalence of anger and its implications in education may be understood in the following two examples. The first example is taken from an ongoing ethnographic study on socially just teaching in Cyprus and concerns the role of emotions in teaching for social justice. Here, I share and analyze a short excerpt of an interview with a teacher at a school that has an increasing population of children who do not speak the majority language (i.e., Greek). In this interview, Joanna reflects on the anger she feels toward the educational system and some of her colleagues, because they seem unwilling to do anything to help minority children.

> To tell you the truth, I am angry at the educational system in Cyprus and I consider us [teachers] responsible for this insensitivity towards minority children in our schools. [. . .] I am particularly angry at many of my colleagues who show no interest to educate children to live in a multicultural society, but continue to teach everyone as if this is a purely Greek society. It seems to me that the "attitude" of the Cypriot educational system, if such a thing exists, is completely apathetic and indifferent to minority children, and this is reflected on many teachers' attitudes as well. I consider this a disgrace for a European country. [. . .] At the same time, one has to consider that the vast social and cultural changes taking place in the last few years, have caught us by surprise. So perhaps I shouldn't be so angry at my colleagues; we all need help in finding ways to address the needs of minority children. We all need to adjust to these new realities. I am trying to do many things but they [other teachers] don't seem to respond positively and I resent that.

Joanna's reflection shows the ambivalent anger that is a part of a politics of struggle taking place in her school community. She struggles to understand her own anger as well as the anger against the educational system and her colleagues. As she emphasizes, there is a lot of work to be done to sensitize teachers on issues of minority and multicultural education. She does not take a monolithic or moralistic position but instead she somehow justifies her colleagues' attitudes. However, she does express her resentment that her efforts so far seem to fail.

The second example builds on the example analyzed in chapter 5, based on Berlak's (2004) efforts to sensitize her students (preservice teachers) to issues of social justice. Reiterating from chapter 5, Berlak invited an African American elementary school teacher (named Sekani) to visit her teacher education class and give a presentation about teaching African American children. This visit involved a role-play situation. Jim, a young white man, volunteered to be the teacher. According to the scenario, there was an immense chaos caused by disruptive children (played by Jim's classmates), and Jim seemed to have a hard time managing the situation. After things continued to be out of control, Sekani terminated the role-play and conducted a debriefing of the role-play. During this debriefing, Jim and Sekani had a heated exchange that made some students feel upset and angry but left many others surprised and shocked in unexpected ways. Their exchange ended as follows:

> Jim: I don't appreciate your comments about rednecks; some of my best friends are rednecks.
> Sekani: Then you may want to reconsider working in a predominantly Black environment. None of us are too keen on YOUR friends.
>
> (Berlak, 2004, p. 128)

Berlak describes how this particular incident created a profound change in her class. As she writes, "Whereas, before Sekani's visit, we had discussed racism in tones we might have used to talk about the weather, afterward virtually everyone was emotionally as well as analytically engaged. Sekani's visit had inadvertently unearthed residual veins of racism and provoked us as a class to confront them" (p. 131).

This incident provides powerful indications about the ambivalence as well as the politicization of anger in education. Anger is ambivalent here because it is a part of a long struggle that constitutes and is constituted by subjects and their relations. Sekani's anger became defined as aggressive and offensive by some students not because it had broken some mutually agreed social rules (e.g., about how class visitors should treat their audience), but because it threatened the taken-for-granted power relations. Students read Sekani's actions as lack of respect for them; therefore, the negotiation of anger between Sekani and the students was about the surprising and moving production and resistance of relations of power and respect (Holmes, 2004a). This ambivalence was acknowledged in class discussions and student journals because it produced conflicting and problematized modes of thinking and feeling. As a result, many students experienced a shift in how they felt, from anger to appreciation, their accusations turned into understanding.

It is important to acknowledge that the effects of moral anger in education, or elsewhere, can be ambivalent partly because anger is so often overwhelming and messy. But angriness can indeed be subversive of power relations and subjectification processes in education, at both a personal and a collective level, as both of the aforementioned examples indicate. An analysis of the politicization of anger in education has to find the way that the personal is linked to the collective: how angry individual feelings about injustice are linked to the social contexts in which these feelings are heard or silenced. Ambivalent anger discourse presents some powerful possibilities. As Bauman writes, "Ambivalence is the limit to the power of the powerful. For the same reason, it is the freedom of the powerless" (1991, p. 179). Again, the question is: What can we (as educators) learn from ambivalent anger and how can we use this knowledge constructively to deal with the politics of trauma involved?

An Educational Politics of Anger?

As I have argued in previous chapters, educational politics and emotions are closely intertwined. Both education and emotions are political manifestations of various kinds of resistances and transformations. The political aspects of emotions, and especially anger, can be better understood in education, if the ambivalence of emotions (Stephen, 2000) is taken into consideration. In particular, the *feeling power* (Boler, 1999) of anger has an important moving role in education in the sense of cultivating individual and collective political consciousness and social resistance to injustices. In this sense, it is acknowledged what has been always the case, in other words, that anger is politicized in education, enabling new perspectives and understandings of oppressive conditions that have been taken for granted in the past.

In the last section of this chapter, I want to suggest that the potential of anger in enabling resistance needs to be recognized in education, although anger is not inevitably emancipatory. Recognizing this could allow educators to understand both the possibilities and limitations that affect individual and collective aspirations about social justice. Listening to students' and teachers' anger about injustices can be constructive, and anger can constitute a central part of challenges to any form of domination. Anger may call attention to a group's demand for respect and recognition, but also highlight inequalities more generally (Holmes, 2004a). The expressions of anger in the two examples I have shared earlier challenge previous assumptions and produce more complex understandings of the events.

In practical terms, expressions of anger are usually dismissed early on because there is a real concern that its escalations may lead to violence and physical injury. However, not all anger is the same. Learning to listening to anger at injustices may lead to constructive classroom dialogue about the fairness of power relations in the society. When this anger is systematically repressed, it is more likely that its repression will lead to resentment, rage, and violence. The threat that this kind of anger poses to schools implies that there are strong cultural norms that seek to suppress the expression of anger and along with it the dialogue about injustice. This dialogue—confrontational or exploratory—is important, because anger may be a powerful source of transformation. The domestication of anger through moral righteousness, care, silence, and technique (Lyman, 2004) is a popular approach of many emotional literacy curricula; however, the ideologies that govern this domestication need to be questioned.

Schools have a lot to gain from producing cultures that recognize and respect the importance of angry speech in addressing injustice, especially if injustice is linked to the perpetuation of trauma narratives (see chapter 6). Individual emotional management of anger is not the answer to conflict resolution and peace when there are social and economic injustices. Reducing the issue to an externally imposed form of emotional management delegitimizes the voice of those who are angry at injustice. Thus, the first step in using anger as a mode of thoughtful analysis, as Grasso (2002) suggests, is to understand where anger is coming from and what power relations are involved. Failing to listen to anger jeopardizes constructive dialogue not only within schools but also in society.

Using anger as a mode of analysis in classrooms can open possibilities to transform how educators and students think about social responsibilities. If our goal, as educators, is to sensitize students, then rethinking the value of politicized anger in the classroom may inspire in students and teachers a sense of *vigilance* (Levinas, 1987b) and social responsibility. Vigilance is what Levinas suggests as becoming a witness to the Other's suffering (see also Sontag, 2003). As he explains, vigilance is "a wakefulness," "a perpetual duty . . . and effort that can never slumber" (Levinas & Kearney, 1986, p. 30). Vigilance is a form of humility that implies that one is being open to an unsettlement—for example, learning from listening to the Other's moral anger, that is, his or her anger against perceived injustices. Thus, it implies being attentive to how teachers and students hear one another's anger and respond to it. Discussing Levinas's notion of vigilance, Oliver writes that "vigilance is necessary to recognize the unrecognizable in the process of witnessing itself. To demand vigilance is to demand infinite analysis through ongoing performance, elaboration, and interpretation" (2000, p. 46).

Educators can benefit from considering how their vigilance could be directed at establishing constructive dialogue with *angry* students in a process that does not dismiss anger but moves all of them toward new considerations of their relations.

In this chapter, I have shown the constructive power of anger discourse in social justice education rather than the constraining conditions of defining it in merely negative terms. Social justice education can be enriched, if anger is taken into account as having an ambivalent and moving role in educating critical individuals. Feminist and poststructuralist writings teach us that anger is not inherently emancipatory, but it is likely to be subversive and as such can be used in educational contexts as well. Accepting anger as a necessary part of disputes embedded in particular power relations can make a difference by moving teachers and students emotionally and politically toward respectful relations with others.

Anger *can* encourage meaningful communication in the classroom, given that it is not treated simply as disruptive of dialogue, even if it may initially seem hostile and aggressive. If anger is expressed directly, there is more scope for teachers to engage productively, but schools are repositories for a great deal of displaced and repressed anger that results in disruption and disengagement from learning. There are clearly issues around providing a safe space for negative emotions in which conditions of trust are key. Students need to have a certain level of trust if they are to voice anger directly. As Holmes (2004b) writes, whether anger "offers salvation or disaster remains dependent on when and how it occurs and the responses people make" (p. 131). The challenge for educators in creating opportunities for responding to injustices and traumatic experiences is to help students experience anger as a positive, creative source of personal and political insight.

CHAPTER 8

DANGEROUS MEMORIES AND THE FORCE OF SUBVERSIVE EXPERIENCES (WITH ZVI BEKERMAN)

> All memory is individual, unreproducible—it dies with each person. What is called collective memory is not a remembering but a stipulating: that *this* is important, and this is the story about how it happened, with the pictures that lock the story in our minds. (Sontag, 2003, pp. 76–77, original emphasis)

> The duty to remember is a duty to teach, whereas the duty to forget is a duty to go beyond anger and hatred. (Ricoeur, 1999, p. 11)

> The past is never dead. It's not even past. (Faulkner, 1951, p. 80)

Historical events—such as the Holocaust, the Vietnam War, World Wars I and II—unavoidably affect the social and cultural sphere of the communities and nations involved. In the past three decades, there have been several attempts to deal with past historical traumas through people's *memories.* Memory is a central issue in contemporary understandings of what it means to do history. Recently, the term *collective memory* (Halbwachs, 1950/1992), written at the beginning of the twentieth century, has been rediscovered and reinterpreted by historians (Santos, 2001). Halbwachs used the term to represent the past within social imagination; in other words, collective memories are understood as collectively shared representations of the past. However, many historians criticized Halbwachs's structural (Durkheimian) analysis and determined anti-individualism (Kansteiner, 2002).[1] Although the alternative terms proposed—such as *social memory* (Fentress & Wickham, 1992) and *collective remembrance*

(Winter & Sivan, 1999), along with other terms such as *national memory, public memory,* and *personal memory* (see Bodnar, 1992)—attempt to overcome the dichotomies between past and present, on the one hand, and individual and social, on the other, it is extremely hard to do so. It is not surprising, then, as Santos (2001) asserts, that memory has become a major issue that is deeply associated with social identity, nation building, ideology, and citizenship.

One of the tensions in the efforts to overcome the classical dilemma between past and present, and individual and social, is whether individuals have a right to *forget* past traumas of their communities in order to construct new, antiessentialist identities that are not locked in past (group) identities, if there is to be reconciliation. "Should we remember? Is it good, is it healthy, to do so? Is it better to forget and move on?" asks Bourguignon (2005, p. 64). This debate forces educators to confront many haunting issues, not the least of which is the relationship between education, memory, and history. The debate is not just about *whether* children should be taught to remember the past, but also about *how* the past is interpreted (Streich, 2002). Given that historical legacies and memories embedded in collective identities cannot be simply wished away, and past historical traumas continue to shape identities and structures in the present (Booth, 1999), this question is redefined. The issue is not whether forgetting is necessary or desirable through history education—forgetting does not imply amnesia here. Rather, the question is whether educators can use past historical traumas to resocialize children in a manner that is not locked into predefined scripts and collective memories (Hill, 2000).

A central concept in this final chapter is an understanding of memory as *dangerous,* that is, as disruptive to the status quo, which is the hegemonic culture of strengthening and perpetuating existing group-based identities.[2] Group-based identities are usually essentialized, static, and tribalistic, because they are built on the notion of separating us (the good) from them (the bad) (Hill, 2000). However, dangerous memories are potentially subversive to those identities and may create new narratives and identities that do not retain essentialism. Needless to say, there is not a particular kind or source of memory that is dangerous per se (Ostovich, 2005). The *danger* is in the practice of re-membering the past in new ways that are disruptive to taken-for-granted assumptions about a group's identity; such ways establish new understandings of personal and collective identities that enable solidarity (Dyson, 1994).

A dangerous memory, for instance, is a memory that breaks through what is assumed to be *true*—for example, that Israelis and Palestinians have always been and will always be enemies—and exposes events and experiences that reveal new and *dangerous* insights for the present situation. A dangerous memory enables solidarity by not appealing to past understandings of identity or ideological narratives but by recognizing shared historical experiences as well as the heterogeneity of those experiences (Hill, 2000). Such memories are also dangerous to national/state educational attempts to subsume the past into unified narratives that further establish traditional dichotomies between us and them. But *how* can (and cannot) the idea of dangerous memories create openings in places where hegemonic powers work tremendously hard to sustain powerful traumatic memories (such as the Holocaust, for example)? In other words, what is at stake in dangerous memories—what is opened up, what is lost—is highly important to examine.

The purpose of this chapter is to sketch a response to this question by exploring the meanings and implications of (dangerous) memories in two different sites of past traumatic memories: one in Israel and the other in Cyprus. Our effort is to outline some insights from this endeavor—insights that may help educators recognize the potential of dangerous memories to ease pain and offer critical hope. However, the premise on which this chapter rests—that is, the importance of raising students' awareness and criticality on the role of memory in their own lives and the lives of those around them—is not new; that premise is not the most important contribution of this chapter. The more important contribution is the analysis of the radically different meanings of dangerous memories in the two sites we explore, and the implications of the complexities in the workings of dangerous memories for educational settings; such meanings and implications are exposed without being locked into perceptions that memories simply "control" individuals or that the political dynamics seek to control individuals by shaping their memories.

First, a discussion on memory, history, and identity sets the ground for discussing the meaning and significance of dangerous memories in the history curriculum. Next, we narrate two stories from our longitudinal ethnographic studies on trauma and memory in Israel and Cyprus; these stories are interpreted through the lens of dangerous memories and their workings in relation to the hegemonic powers that aim to sustain collective memories. The two different stories suggest that collective memories of historical trauma are not simply *transmitted* in any simple way down the generations (see Sorabji, 2006), although there are powerful workings that support this transmission. Rather, there seems to be much ambivalence in the workings of memories that under some circumstances may create openings for new identities. The final section discusses the possibilities of developing a *pedagogy of dangerous memories* by highlighting educational implications that focus on the notion of creating new solidarities without forgetting past traumas. This last section employs dangerous memories as a critical category for pedagogy in the context of our general concern about the implications of memory, history, and identity in educational contexts.

MEMORY, HISTORY, AND IDENTITY

As mentioned, Halbwachs's work on collective memory has been widely revived in recent years. Yet the concept has undergone much criticism. Conway (2003), for example, argues that the term is elusive for it is difficult to delineate the relationship between what is actualized at an individual level and what constitutes collective memory; others consider it so vague as to lack merit (Gedi & Elam, 1996; Olick, 1999) for it sensitizes people to the influence of the social and political context on the formation of memory (Kansteiner, 2002; Middleton & Edwards, 1990). Nonetheless, Conway (2003) observes that there are two prominent meanings in the literature: First, collective memory refers to people's shared recollections of past events; and second, the term refers to past memories that are embodied in *technologies of memory* (mass-media products) that help transmit them to future generations. In both cases, memory is a kind of (individual-social) performance in which the act of remembering is the intertwining of both the individual's will and social determinations (Olick, 1999; Santos, 2001).

The notion of memory has taken its place now as a leading term in cultural history (Confino, 1997). But what is the relation between memory and history? Kansteiner (2002) argues that memory's relation to history remains one of the interesting challenges in the field for there are good reasons to question a clear epistemological divide between history and memory in spite of the acknowledged fact that people remember selectively (Epstein & Lefkovitz, 2001). As Burke (1989) remarked: "Neither memories nor histories seem objective any longer. In both cases we are learning to take account of conscious or unconscious selection, interpretation and distortion. In both cases this selection, interpretation and distortion is socially conditioned" (p. 98).

The French historian Pierre Nora (1989) helped revive the debate on collective memory by arguing that memory is no longer a servant of history but on par with history. Nora indicates that memory is always subject to the dialectic of remembering and forgetting (Kansteiner,

2002). Yet other scholars emphasize the necessary dialogue between history and memory (Siegel, 2002) denying their rivalry and underlying their complementarity when relating to the past (Assmann, 2006; Siegel, 2002). The point of memories, then, is less to tell us what *exactly* happened than what it *felt* like to experience an event.

Consequently, memory plays a major role in structuring national identity (Kansteiner, 2002) and sustaining a sense of self in and through the communities in which individuals belong and relate to others (Epstein, 2001; Middleton & Edwards, 1990). The connection between memory and identity, then, raises two important issues: First, it highlights the political and emotional value of collective memories because past representations are preserved through social and ideological practices such as commemoration sites and rituals; second, the connection between memory and identity suggests that memory is created in interaction between and among people in social and political contexts (Conway, 2003; Middleton & Edwards, 1990; Olick & Robbins, 1998; Zerubavel, 1996). To say that memory is socially and politically situated is not to deny that it has an emotional dimension unique to each individual; on the contrary, memory is both individual and collective and is constituted in affective economies of shared remembering (Ahmed, 2004).

What gets defined as the *official* memory, therefore, reflects the power of certain groups and ideologies in society to define the past according to their interests, often by silencing alternative and competing memory discourses (Conway, 2003; Epstein, 2001; Middleton & Edwards, 1990). This is particularly obvious in historical narratives taught in schools when such narratives provide a framework through which children make sense of and lay claim to a national collective memory (Davies, 2004; Siegel, 2002). History curricula implore students to remember the nation's glories, leaders, and warriors through practices that aim at establishing a historical consciousness that "aligns forgetting with evil forces" (Eppert, 2003, p. 186) that threaten to destroy the nation's identity and its very existence.

While students and teachers may view state-sanctioned curricula and pedagogies as simply *the truth* about past collective memories, such practices are *technologies* (Foucault, 2003) formed and circulated to promote particular ideological practices. These technologies select and organize what can be legitimately known about the nation-state and its supposedly glorious character, and legitimize both the existence and the governance of the nation-state as normal and unproblematic (Montgomery, 2005) creating a social world of two opposing identity categories of us and them (Davies, 2004). However, students and teachers are not dopes answering the mandates of the *politics of memory* (Simon, 2005; Todorov, 2003); instead, a sense of rupture with official historical narratives and essentialized identities may be grounded in the notion of *dangerous memories* for they challenge those educational assumptions that suggest that "transmitted memories" are endlessly powerful.

Dangerous Memories

Dangerous memories are not a particular kind or function of memory that can be isolated and defined, points out Ostovich (2002, 2005); rather, they are "a *disruptive* practice of and from memory" (2002, p. 239, emphasis added). Any memory can become dangerous when it resists the prevailing historical narratives. Consequently, dangerous memories are neither simply individual nor collective but *political* in the sense that they involve power relations, revealing the patterns of violence and suffering at work. The concept of dangerous memories has been widely cited in education discourse in the 1990s, especially through the writings of Giroux (e.g., 1992, 1997). Here we want to build on this concept and argue that dangerous memories may constitute new affective economies that inspire solidarity through the memory of suffering. These ideas are discussed below, offering a ground of critical understanding in the practice of memory through gesturing to some of the conditions that might be necessary for such a solidarity to be actualized.

The German political theologian Johann Baptist Metz has offered a detailed theorization of the notion of dangerous memories, building on some of Benjamin's work. Deeply aware of the connections among memories and identities, Metz offers a profound critique of taken-for-granted memories and the mechanisms through which such memories encourage feelings of fear, anger, and revenge toward others. As he states: "Identity is formed when memories are aroused" (1980, p. 66).

Metz (1972, 1980) argues that there are two categories of memories. In the first category, memory is simply the recollection of the past, but it is a *selective* memory; it remembers the past only from the perspective of self, thus perpetuating the status quo. In the second category, memory interrupts fixed historical narratives by acknowledging the powerfulness of human suffering. All humans as subjects are located in suffering; thus, through the memories of suffering—that is, *dangerous memories*—the taken-for-granted narratives are interrupted. As he writes:

> There are memories in which earlier experiences break through the center-point of our lives and reveal new and dangerous insights for our present. They . . . show up the banality of our supposed "realism." They break through the canon of all that is taken as self-evident . . . [and] subvert our structures of plausibility. Such memories are like dangerous and incalculable visitants from the past. They are memories we have to take into account: memories, as it were, with future content. (1972, p. 15)

Dangerous memories are disruptive because they call for a solidarity with others on the basis of common human suffering. These kinds of disruptions come as dangerous memories when we remember events of the past that question our consciences and assumed horizons; *dangerous,* then, takes the meaning of challenging, critical, and hopeful while propelling individual and collective consciousness into a new process of narrativization. Reclaiming forgotten connections with others involves acts of compassion, self-criticality, and resistance to the status quo.

As we are trapped in egotistic and ethnocentric mentalities, dangerous memories interrupt our endless cycle of selfishness and open up our eyes to the suffering of others as well. As Metz notes in the context of violence and hatred in former Yugoslavia,

> The memory of suffering became a shroud for the whole nation and a stranglehold on any attempt at interethnic rapprochement. Here a particular people have remembered only their own suffering, and so this purely self-regarding *memoria passionis* became not an organ of understanding and peace, but a source of hostility, hatred and violence. (1999, p. 230)

Following the spirit of the political theology of Metz means that the patterns of past violence and hatred may be subverted and solidarity with former enemies can be inspired through the memory of common suffering with others. What constitutes such a solidarity is empathizing with the sufferings of others, grounded in *feeling with* the others' suffering (see chapter 2). Obviously, the cultural circulation of memories of others' suffering does not always result in formations of solidarity. As it has been argued in previous chapters, relating to others' trauma without sentimentalizing suffering is not an easy task (see chapter 1).

Metz's vision for a renewed solidarity that deeply empathizes with the sufferings of others offers a profound critique of business as usual and challenges us to *bear witness* to the Other (Oliver, 2001). In other words, dangerous memories constitute a *praxis* of memory, that is, a practice grounded in solidarity with the marginalized others. As Metz writes: "It is in . . . solidarity that memory and narrative . . . acquire their mystical and political praxis" (1980, p. 229). This idea suggests a radical questioning of the collective memories that are taken for granted. It is this solidarity that offers a way out of the sanitized narratives of selective memories within which the narrow us is barricaded against the fearful them (Ashley, 1998). On a personal level, this solidarity requires a constant openness and criticality to one's self and transformation and a willingness to recognize our connections to another's suffering—through attention to their memories of suffering such as listening to their stories and working with them to alleviate suffering. In this way, we begin to see ourselves as interdependent and vulnerable to injustices.

Welch's (1985) book, *Communities of Resistance and Solidarity*, which examines the role of communities in both fostering liberation and sustaining oppression, builds on Metz's work arguing that dangerous memories fund a community's sense of dignity because they empower those who challenge oppression and injustice. For Welch, dangerous memories are a people's history of resistance and struggle in the face of oppression; they bring to the surface stories of injustice and mark how recognizing past wrongs may endanger the continued acceptance of injustice and thus propel people to acts of resistance. Welch writes that communities of resistance may correct such injustices; we need such memories, she argues, if we are to find and sustain the strength, resilience, courage, and hope required for resisting the forces of evil. Giroux (1997), who introduced Welch's thinking in education, suggests that

> transformative intellectuals need to begin with a recognition of those manifestations of suffering that constitute historical memory, as well as the immediate conditions of oppression. The pedagogical rationality at work here is one that defines radical educators as bearers of "dangerous memory," intellectuals who keep alive the memory of human suffering along with the forms of knowledge and struggles in which such suffering was shaped and contested. (p. 105)

Therefore, according to Giroux, dangerous memory has two dimensions "that of hope and that of suffering. . . ." (ibid.).

Finally, the praxis of dangerous memories can also be described, after Benjamin (1968), as *anamnestic solidarity*. Anamnestic solidarity with others—past victims—forbids the unreflective and instrumental appropriation of prevailing historical narratives and demands instead a self-critical and committed reappraisal of collective memories (Duvenage, 1999; Pensky, 1989). As Pensky argues, this criticality means abandoning conventional forms of collective memory and identity and anchoring relationality with others in telling the stories of those who suffered. But how easy is it to abandon conventional forms of collective memory and identity and engage in a praxis of memory grounded in solidarity with the Other's suffering? Is it possible that essentialist meanings of memory co-opted by the national state negate efforts to engage in anamnestic solidarity? In the following part of the chapter, we focus on two examples of the workings of (dangerous) memories in two conflict-ridden areas, Israel and Cyprus, in which collective memories about past suffering are both abundant and powerful.[3]

MANAGING MEMORIES OF HISTORICAL TRAUMA: NARRATIVES OF PAIN, NARRATIVES OF HOPE

The two stories we narrate here—one story written by Zvi and the other by Michalinos—were based on ethnographic data that each of us has collected over the last 5 years. These data are part of larger projects that explore, among other things, the role of memory, trauma, and identity in the narratives of teachers and students in the conflict-ridden areas we live. Our stories were compiled from analyses of interviews, observations, workshops with teachers, and other data collected.[4] These stories were written with the purpose of showing how teachers (in the first example) and students (in the second example) work with their memories to gain an understanding of the past. By writing these stories, we leave numerous others unwritten; however, our interest in telling these stories is to gain a richer understanding of what makes memories so difficult to overcome and whether there are openings for transformation in the ways that collective memories are remembered in different social and political contexts.

Yoni and Painful Memories (Written by Zvi)

I was lucky to be able to work together with the teachers at the integrated bilingual schools in Israel within the context of workshops I conducted a couple of years ago. Lucky I was indeed but the experience was not necessarily encouraging. The second three-day workshop with the teachers opened with an activity in which they were asked to think about issues they might in the past have found difficult to raise in the presence of representatives of the other group. At first, they were asked to write down their thoughts, and when this stage was completed, the Jewish teachers were the first to offer their accounts.

Yoni was the first to speak. Yoni is today one of the few Jewish teachers at the bilingual schools who is fluent in Arabic and is also a declared "conscientious objector" who does not serve in the military reserves, thus positioning himself in the far left liberal wing of the Israeli political spectrum. As did five of the six Jewish teachers who took turns in the dialogue, Yoni mentions military service and terrorism as central issues he has difficulty discussing with the Palestinian group. He says, "It gets me angry when a Palestinian who doesn't support terrorism, doesn't condemn it. He says 'he [the terrorist] is a part of my people.' Now, when I'm with friends and family and I know that they serve in the Territories [conquered by Israel in the 1967 war], I don't necessarily condemn it." It seems that from Yoni's perspective, Israeli Palestinians, in spite of their present situation as a peripheral and subjugated minority in the state of Israel, should condemn terrorism as outright evil as he does.

When it is the turn of the Palestinian teachers to hold the floor and offer their accounts, Nadia says: "There are things that are reality, and that have to be heard. They [the Jews] shouldn't expect to hear just the things that are acceptable to them." Yoni reacts acknowledging that the traditional Jewish-Palestinian asymmetry seems to be reversed in the dialogue: "I want to add that sometimes part of my difficulty is the feeling of having to (or being expected to) apologize—as if I have to justify myself. That's something I don't feel from the other side [i.e., the Palestinians do not see themselves needing to apologize]." Ibtisam is fast to pick up on this asymmetry reversal and responds to Yoni, "I had a feeling that you do go through a process [you better understand the problems of the Palestinians], but it's contingent. As soon as the other side doesn't meet your expectations [to apologize] you're disappointed. And I wonder what it is we should apologize about."

Ela, a Jewish teacher, follows with a rather long statement in which a possible (and forbidden in Israeli hegemonic terms) comparison between the plight of the Palestinians and that of Jews in the Holocaust is heard: "I feel that there are two historical narratives and that it blends in with personal feelings. I know that the Palestinians take more of an interest in our narratives than we do in theirs. Sixty years after Auschwitz, there's now discussion of Americans and English knowing what was going on and why they didn't bomb the railroad tracks. It's said that they didn't know what they'd do with another million Jews. And I see how important it is that the Palestinians know my story. . . . When Manal spoke I thought 'speak. I want to hear.' It's not a question of agreement or not. It's not 'wow she's also against the Palestinian attacks.' I want to know more." Ibtisam follows with a question, "Can I ask something? Do you mean to do something with the knowledge?" and after a short exchange with Yoni adds, "You said it angers you if a Palestinian doesn't condemn terrorist attacks. What angers you? I'm not a saint. I'm dying of fear. I haven't done anything. What should I apologize about?"

The teachers in a rather short time have been able to cover the totality of the Palestinian-Jewish conflict as if their participation in the joint educational experiment has made no difference other than the fact that issues are put on the table seemingly, though painfully, with ease. The problems raised point in the direction of a "fearful symmetry"; each argument raised on one side has been countered on the other. The conflict is attached to the existence of two narratives (symmetrical and oppositional).

Jews were the ones who mostly feared speaking openly about historical and present issues in the mix group with Palestinians; they felt that if they became critical their criticism might serve as "ammunition" to the Palestinian side. I was surprised by this fact. Jews, even the ones involved in a bilingual multicultural educational initiative, expressed traditional Zionist perspectives as if their working in the context of the integrated initiative had not influenced them (almost) at all. Though critical of some of the behaviors of the Israeli armed forces, for the most part, they justify their activities both in the past, during the war of 1948, and in the present intifada.

What most surprised me, however, was that even the critical insights presented by the scholars teaching at the workshop, which clearly pointed at a wide agreement among historians about the historical "facts" and related to most of the official Jewish Zionist history presented at schools regarding the events of 1948 as mythical, did not seem to allow for much change in the positions of the teachers.

Yoni's anger and discomfort, on the one hand, and his liberal-left views, on the other, testify to his ambivalence to come to terms with the memories about what happened in the past. Yoni and the other Jewish teachers attached themselves to a historical deterministic perspective (what was will also be—Jews have been persecuted and will be persecuted, if not careful enough) that allowed little space for optimism. They also expect symmetry on a variety of levels. They wanted, as they easily did, the Palestinians to condemn "terrorism" unconditionally; terrorism as understood from the Jewish perspective, thus in a way expecting Palestinians to recognize the 1948 war as the Archimedean point from which the discussion should start and, following this reasoning expected Palestinians to invest their effort as

Jews did after the Holocaust (their own Archimedean point of departure) in reconstructing their community instead of continuing to struggle for recognition as a people. In a sense, they were asking them to forget the Nakba while forgetting the Holocaust was the one thing they were not ready to do. Palestinians contested the Jewish expectation for both symmetry and asymmetry: symmetry regarding the present political situation in Israel and its conflict with the Palestinian people; asymmetry regarding the nonequivalence between the way Jews suffered in the Holocaust and the way Palestinians suffer at the hands of Jews.

Odd as it may sound, one has to search for openings in Yoni's sense of ambivalence in order to advance a radically different understanding of memories. Clearly, this is not easy to do, because collective memories seem to resist efforts to disrupt the status quo. Although Palestinians' suffering is put forward as a potentially subversive element, Yoni's fixed memories of the past subsume this suffering into a unified narrative of a Jewish-only suffering. In other words, memories of Palestinian suffering are treated asymmetrically compared to memories of Jewish suffering. It is in this sense that memories of the Palestinian suffering as well as the Palestinians' refusal to condemn terrorism constitute *dangerous* memories for Yoni's attempts to subsume the past under his predefined structures of collective identity. The mechanisms of memory in the national culture of Israel prove to be very powerful to break through them. Nevertheless, dangerous memories here are translated into challenging and discomforting memories, despite the fact that there is not much space left for transformation.

Anna and Hopeful Beginnings (Written by Michalinos)

Anna took two of my teacher education courses at a private university—as a sophomore and as a senior—so I got to know her well. She is 19 years old and her parents come from the occupied territories in north Cyprus. She was born after the war of 1974 and brought up in a house full of pictures from her parents' village; therefore, she always felt strong ties to the "occupied land" [katehomeni gi] in spite of the fact that she never visited this land until recently. Her parents refuse to "show passports, as if they are visitors in their own land," she said. After visits to and from the north were allowed in April of 2003, many Greek Cypriots and Turkish Cypriots crossed the checkpoints to the other half of their country for the first time after almost 30 years. But still many Greek Cypriots—including Anna's parents—refuse to visit their villages and prefer to live in nostalgia until the day their land is "liberated." Most of those who visit the occupied territories say that "everything has changed a lot" and "they [the Turks/Turkish-Cypriots; no distinction is made] haven't cared for our houses and our land."

When Anna announced that she had decided to visit "the other side," her parents reacted angrily. "I never thought they would be so deeply hurt," she told me. "They began crying and said, 'What are you going to do? Are you a traitor? Have you forgotten how we raised you?' "

As Anna began to narrate her story, the memory of this particular emotional experience with her parents was obviously too painful for her. I wondered, though, What made this tremendous shift in her life, given that she had grown up believing that "liberation struggle" is the only route to return [epistrofi] to the homeland of her parents?

Like many Greek Cypriots, Anna grew up learning in school how "barbaric" and "evil" the Turks were. "They have done this, they have done that; that's how I grew up in school," she said, and continued: "My parents had to flee from their village to save their lives; they became refugees in their own country. Our houses, churches, and historical monuments were looted and destroyed. There are still hundreds of missing persons [agnooumenoi]. I learned not to forget all these; everyone taught us that we can't forget. . . . The slogan 'Den Xechno kai Agonizomai' [I don't forget and I struggle] was quite prominent when I was in elementary school."

In May of 2005, Anna met a twenty-year-old Turkish Cypriot, Ahmed, who happened to be drinking coffee at the university cafeteria. This was the first time she met "one of them." Ahmed was also a student at the university; he was among the first Turkish-Cypriot students to cross the checkpoints and come to this particular (English-speaking) university in the south. To Anna's big surprise, Ahmed lived in the same village that her parents came from. Ahmed was also surprised, when he heard Anna describing the village—its streets, its houses, the Greek church, the old mill—as if she was there the day before. Over the years, Anna created an almost perfect map of the village in her imagination—except the addition of a few recent houses—completely out of her parents' memories and the pictures she had seen. Ahmed invited her to visit her parents' village. Without an instant of hesitation, without knowing why, just like that, Anna accepted the invitation.

A few days later, Anna and Ahmed were walking in the streets of the village. At some point, they ran into an old man in an armchair, overseeing their conversation [they were speaking in English, as nobody spoke each other's language]. Anna and Ahmed greeted the old man and introduced themselves. As soon as Anna pronounced her first and last name, the old man asked her [in Greek]: "Are you the daughter of Andreas and Maroulla?" Anna was shocked. "Yes! But how do you know?" she replied. The old man said: "You look exactly like your mother! You know, we used to be neighbors with your parents. Let me take you to your house." Before Anna realized what had happened, she was in the backyard of her parents' house. Indeed, she recognized it from the pictures and the stories her parents had been telling her all these years. On each side of the pavement that was leading to the main entrance, there were two olive trees and a rose garden that seemed to be well taken care of. Anna was in tears by now. She was barely able to whisper, "Can we enter the house?" The old man replied immediately: "Of course! Nobody lives in it. I come occasionally to water the flowers. Tell your father that the key is under the carpet, just as he left it. Please tell him that Bulent takes care of his house until you come back. Don't forget to tell him!"

As Anna narrates the story to me, she picks up from her bag a beautiful dried red rose. "From our garden," she says smiling. "When I told my parents what happened, they were crying for three days! But they wanted to learn every little detail about the house, the garden, the village. And then they told me stories about their Turkish-Cypriot neighbor, Bulent—for the first time! I don't know why they never mentioned him before. . . . Now they consider visiting the village but they are too stubborn to admit it. I understand how painful it is for them. I understand. . . ."

Anna tries to explain the epiphany moment that changed everything: her transformed perspective about Turkish Cypriots and how she realized that they had also been refugees in their own country, they too had missing persons and lost loved ones in the war—contrary to everything she had learned in school. "You know what?" she told me, "I don't know how to explain this. I don't know how it happened. It just did. What matters to me is that I don't see them [Turkish-Cypriots] any more as the monsters described to us in school. They are human beings, just like us. . . . It's not that I am not struggling to accept their own suffering. Oh, believe me, I struggle very hard! It's hard to escape being socialized in conflict for so many years. But meeting Ahmed and experiencing his generosity makes me want to learn more about their own sufferings. . . ."

Perhaps part of what Anna is expressing is the inability to believe that memories can convey such different and ambivalent feelings.

When past memories take a new life and recent ones seem impossible as part of a world that is taken for granted, then remembering and forgetting become something that transcends this world. Perhaps this is why there seems to be some kind of discontinuity, an interruption of things as usual. What seems to be most surprising is that despite the painful memories with which Anna grew up for most of her life, there was still some room left for navigating her memories toward a different path.

Anna's past memories are contested by recent ones in which her experiences with the "enemy" break through the main narrative of her life and reveal new—and *dangerous*—insights. Her memories become dangerous from the moment that they begin to illuminate the questionable nature of past beliefs that were taken for granted. This danger is more obvious in Anna's clash with her parents about visiting their occupied village. Anna's visit seems to unmask the certainty of things that were perceived as self-evident in the past (e.g., her perceptions about Turkish Cypriots). Unavoidably, the newly discovered resistance to collective narratives puts dangerous demands on the present. But does this effort demand "forgetting and moving on from a new foundation . . . ?" as McKnight (2004, p. 150) asks. In other words, are there forms of forgetting that, while not requiring amnesia, lead to new directions that are not locked in past historical categorizations? These are challenging questions, and the concept of dangerous memories does not improve the situation. Forgetting is usually criticized for attacking the nation's foundations or for denying foundational memories (such as the Holocaust, for example). However, is it possible perhaps that some forms of forgetting (compared to others) might provide new insights into solidarity with the Other?

As Ricoeur (2004) and Eppert (2006) argue, it is less an issue of remembrance versus forgetting than it is some forms of remembrance versus other forms of remembrance and some forms of forgetting versus other forms of forgetting. In other words, it is not that the unjust past and the suffering are being forgotten. Rather, it is the anger and the hatred that are being forgotten, so as to enable space for reconciliation. Certainly, this seems to be Anna's case. She is not forgetting her own and her parent's suffering; but through her efforts to remember the suffering of the other side, she has an opportunity to choose to forget a unilateral narrative of history—to forget the lessons of her schooled past that told her she cannot forget but must fight against "them" (must, in other words, remember to stay hating). She has an opportunity to choose to forget her enmity.

TOWARD A PEDAGOGY OF DANGEROUS MEMORIES: STRUGGLING FOR ANAMNESTIC SOLIDARITY IN EDUCATION

Although the two stories we narrated take quite different directions in terms of the meanings and implications of dangerous memories, they both show the ambivalence and the challenges involved in considering the potential of dangerous memories to enrich historical understanding in educational settings. On one hand, the story from Israel suggests that the idea of dangerous memory might be difficult to work in places where hegemonic powers work tremendously hard to sustain past traumas (i.e., the Holocaust); present essentialist meanings about memory and identity so much supported by the national state constitute machineries that seem hard to work against. On the other hand, the story from Cyprus indicates that despite the powerful hegemonic workings of collective memories, there might be openings for dangerous memories to subvert the status quo—not at the collective level but at least locally. Therefore, it becomes interesting to explore the sociopolitical and educational conditions under which dangerous memories can/cannot be facilitated.

First, it is important not to read the two encounters with past traumatic memories in a polarized manner, that is, the Cypriot case demonstrated an opening to dangerous memory while the Israeli case did not. The differences between the stories should first be interpreted in terms of the fact that not only are we speaking about two individuals and no individual trajectories ever can coincide—especially when considering Bakhtin's (1984) principle of noncoincidence that argues that not even A=A—but we are also speaking about individuals experiencing different sociohistorical contexts, both indeed conflictual but nevertheless different. These differences also have to do with the ways the contexts allow for the organization of engagements with past memories, present realities (demographic and other), and future potential expectations. The Cypriot reality has organized a society fully divided, whereas the Israeli context has been organized around the existence of a subjugated indigenous minority and a Jewish majority. From this perspective, while individual encounters might and do occur, they become more or less available and are shaped according to contextual constraihts. For Israeli Jews, the Palestinian presence of the Palestinian citizens of Israel is perceived as a threat, a potential fifth column, a continuous reminder that someone else holds a claim to the land they inhabit, and last but not least, a reminder that within them sit representatives of their feared "enemies," the Palestinians outside the borders of Israel in the Palestinian Authority and other Arab countries. For Greek Cypriots, Turks, or Turkish Cypriots have been less available until recently (when the checkpoints were opened) and yet they have always been perceived as a threat or a reminder of a threat more or less immediate. In both Israel and Cyprus, the conflict is still present though indeed bloodier in Israel and with fewer expectations for its near resolution.

Moreover, to the differences above we should add the immediate context in which the encounters take place. For Greek Cypriots like Anna, it is an encounter in a daily event while the Turkish Cypriot is sitting at a cafeteria in a university; it is two individuals meeting each other in a situation that though accompanied by a *history* and much *memory* allows for multiple directions to develop in the social setting and interaction. The encounter in Israel takes place in an organized formal setting, a training program for teachers in an integrated bilingual (Hebrew-Arabic) school. The themes they are discussing are directed toward the most inner fibers that compose the conflict—the historical opposing narratives that for each are essential to justify their claims. Indeed, other options could have developed, but in educational settings, the core of the official history dissemination seems to allow for little other than what took place. In Cyprus, the encounter with the Other still takes place between almost complete *strangers* so there might be the element of *surprise* when meeting the Other. Anna and Ahmed experienced surprise in the context of their sharing of memories and experiences. The element of *surprise* is probably a good indication that some kind of learning is going on beyond the dissonance created between the abstractions of the enemy-Other and the individuals-in-relation to each other. In the example with Jews and Palestinian teachers, there seem to have been no surprises experienced: The encounter between individuals really takes places between *groups,* because the individuals are caught in the social and official narratives about the Other.

Finally, the comparative consideration of the two stories provides insight into the complex *layering* in the space between the *layers* of witnessing the Other's suffering; layering is the event of dangerous memories taking shape and form across different times and spaces. It is not enough to understand how to place one layer of dangerous memories on top of another; the task is to explore what happens between the layers, in the layering itself. In Israel, memory is grounded on suffering (the suffering of the Jews); it always seems as if the struggle is who suffered more. The liberal Jews are ready to accept that Palestinians suffered but as many say they have many other countries whereas the Jews do not; therefore, Jewish suffering allows the Jews while recognizing Palestinian suffering to disregard it. Our story from Israel is a good example showing the layering of witnessing the Other's suffering. The example from Cyprus, however, shows a different kind of layering concerning the witnessing of the Other's suffering. That is, getting to witnessing the Other's suffering passes first through witnessing the Other's kindness. It is Ahmed's kindness and the loyalty of the Turkish-Cypriot neighbor, Bulent, that enables Anna to open her heart and mind to the Other's suffering. In this case, we have a different kind of layering in the development of dangerous memories and their role in the potential to create some sort of new solidarity.

What, then, are the pedagogical implications of the complexities and nuances of dangerous memories—as the two stories have shown—for engaging students and teachers with the difficult issues of remembrance, forgetting, history, and identity? In other words, what is the potential that dangerous memories may really constitute subversive forces in educational settings, given the hegemonic power of a nation's foundations? Is this something feasible or even desirable? Undoubtedly, to examine aspects of remembrance and forgetting and draw them out of taken-for-granted perceptions into a critical realization of the relationality with others is an immense pedagogical challenge (Simon, 2005; Simon, Rosenberg & Eppert, 2001; Zembylas, 2006b). This examination, as the stories of Yoni and Anna suggest, includes the praxis of dangerous memories, that is, a self-criticality that needs to begin with a critical analysis of historical circumstances and memories, their connection to the present, and the recognition of the Other's suffering. But again, there is not any assurance that when memories become dangerous, then transformation will take place. All by themselves dangerous memories cannot take us too far; as any other symbolic tools, they are always, on their way, somewhere but the *where* is unknown.

It is important to also recognize that the same traumatic memories may produce different outcomes; for the Zionist movement, the Holocaust became the ultimate justification in their claims for an independent state and their positioning as the only victim, while the ultra-orthodox Naturei Karta found in it the return to Zion a rebellion against the divine decree of exile thus exposing themselves to more divine retribution (Boyarin & Boyarin, 1995; Ophir, 2000). Similarly, for the Greek-Orthodox Church and the Greek-Cypriot Right (strongly nationalist), the Turkish invasion was another example indicating that Turkey is an inherently barbaric and expansionist country and the only available option is resisting this barbarism; in contrast, the Greek-Cypriot left-wingers who had the best relations with Turkish Cypriots in the past and sometimes found themselves victims of right–wing nationalist violence adopted a discourse of rapprochement with Turkish Cypriots (Constantinou, 2001; Papadakis, 1998). The workings of essentialist meanings about collective memories are so deeply ingrained in a nation's foundations and are cited so many times throughout many levels of society (Kumashiro, 2002) that it is just too difficult to create openings. Yet, as Kumashiro argues, it is possible to *supplement* these citational practices with new meanings that do not necessarily *repeat* essentialist histories. As Anna's story shows, developing a critical awareness of dangerous memories may begin to alter the citational practices that constitute hegemonic beliefs. Nevertheless, what is the usefulness of exposing dangerous memories in educational settings, if there is so much ambivalence involved?

First, the site of dangerous memories is important in terms of establishing a praxis of dangerous memory—that is, creating a set of commitments to and actions that clarify the openings in various forms of remembrance and forgetting. In the stories we narrated, there are opportunities to bring dangerous memories into the classroom or workshops; in other sociopolitical situations that might not be the case. There might be danger, for example, to the careers and lives of participating teachers and students. Anna's situation is enabled by the particularity of the encounter; not only did she have the opportunity to meet the other side, but she was also shown care and respect.[5]

In addition, dangerous memories, our stories show, are unsettling enough for students and teachers to enable *supplementation* through a reworking of one's relationships to the world and others. The results may vary, but nevertheless a reworking of relationality is in progress. For example, both Yoni and Anna had difficulties establishing relationality with the Other. But despite these difficulties an exchange with the Other is being worked out. The ambivalence felt by both protagonists is constitutive of this effort; the end result is unknown. The fact that Anna's narrative leaves us with some hope does not imply that the situation will be necessarily transformed. What is important, in our view, is that teachers and students need to try very hard to escape from remaining prisoners of collective memories. Freeing ourselves from collective memories, however, is not solely the task of formal educational activities. These activities are necessary but are in need of support from multiple social political sites ready to engage in a similar quest. The praxis of remembering suffering is precisely an attempt to formulate the pedagogical conditions of possibility so that students and teachers may disrupt egocentric goals. In general, it is important for teachers and students to move beyond thinking of collective memories as the primary source of information about the past.

Dangerous memories offer an interesting alternative: In stressing the importance of critical understanding in the practice of various forms of remembrance and forgetting, the praxis of remembering that suffering is common might become a medium for opposing an unreflective account of the past. Even in the first story, things are not as gloomy; as soon as the Palestinian teachers pointed out the asymmetry in Yoni's understanding, the memory of terrorism became disruptive. This disruption initiated a conversation that might lead somewhere else. At least this is what is suggested in the second story; Anna picks up the challenge from where Yoni has left it. After her (recent) memories of her parents' village become unsettling, she begins to *see* the other side's suffering. In her effort to deal with everyday life events, Anna responds to her encounter with Ahmed; in this encounter, there are surprising consequences. It is not that the structural constraints are not there, managing to reproduce the system quite successfully; new insights, however, become personal and meaningful—these are the insights that may lead Anna in new directions.

It follows from the disruptive character of dangerous memories that getting *to know* the past does not mean *controlling* it. Instead, *knowing* requires an act that "re-cites and re-sites what one has learned—not only about what happened to others . . . but also (and this is key) what one has learned of and within the disturbances and disruptions inherent in comprehending the substance and significance of these events" (Simon, Eppert, Clamen & Beres, 2001, p. 294). In Simon's (1993) terms, dangerous memories may serve as *insurgent commemoration* or *counter-commemoration* in offering disruptive alternatives. Such an approach takes seriously a nontotalizing view of the past without the naïve acceptance that the past *determines* who we ought to be (Duvenage, 1999). Through Yoni's story it is shown how challenging this process is, because it is really hard to escape from some foundational memories; on the other hand, Anna's story suggests that despite these challenges there is no lack of hopeful openings. In any case, both stories teach us that the praxis of memory requires an anamnestic solidarity with others who have also suffered and thus precludes any simplistic closure based on *our* narratives. The pedagogical implication of anamnestic solidarity in light of dangerous memories is the responsibility to question forms of remembering and forgetting and not set them as polar opposites, but provide spaces for telling heterogeneous stories, despite the seemingly insurmountable difficulties.

In other words, the issue is not about remembrance versus forgetting. A *pedagogy of dangerous memories,* if it may be called as such, neither presupposes nor ends in any polarity between remembrance and forgetting; on the contrary, it presupposes a form of recollecting past suffering (Krapp 2005), thus it takes us to the limits of memory and forgetting, but the recollection of past suffering is not only fixed on "our" suffering. This is why it may be argued that the praxis of memory based on suffering "is a modification of forgetting that does not affect the irrevocable, nor repress its memory" (Krapp, 2005, p. 191). Dangerous memories (as any other introduction of elements into a system) force the system to calibrate again, to restabilize, to readapt, and to reconfirm in order to reach homeostasis—this is the power of new knowledge (in our case dangerous memories). Again, the working with dangerous memories is never a promise for success but they set the system in motion and with motion (not stasis, death) new opportunities are open. These opportunities are the possibility of renegotiating meaning through the repositioning of self in dialogue. In the specific case of Zionist Israel, this could mean the opportunity to move from the position of victim as a cultural construct produced and supported by multiple tools (of state) production to one that allows Palestinians to be seen as victims too. Similarly, in the case of Cyprus this could also mean the opportunity to abandon the immovable categories of Greek Cypriots (orTurkish Cypriots) as the only *victims,* while the historical category of *perpetrators* is always saved for the *other* community. In a sense, therefore, dangerous memories are not about forgetting versus remembrance but constitute instead a widening memory to include the Other's memory. This is an interesting idea to consider—an idea that is oppositional to claims that amnesia may be required to move on. To put this differently, dangerous memories constitute inclusive memories that oppose the dichotomy between forgetting/remembrance. Some forms of forgetting (e.g., hatred) and other forms of remembrance (e.g., the Other's suffering) may be needed to overcome hostility.

That is, even some purposeful forgetting does not eliminate memory, but allows the overcoming of hatred, resentment, and violence. However, the situation is complicated, as the first story shows, when forgetting is used to further establish the existing asymmetries between groups. Nevertheless, it is important to consider Ricoeur's (1999) reminder in the motto with which this chapter begins that, "the duty to remember is a duty to teach, whereas the duty to forget is a duty to go beyond anger and hatred" (p. 11). Forgetting, adds Eppert (2003, 2006), is not only bound up with obligation, but also with an obligation implicated in reconciliation. In both stories, forgetting past suffering is almost impossible; however, in Anna's story the window left provides a different interpretation for the "responsibility" to remember the past. Here Ricoeur and Eppert move around forgetting and thus challenge us not to fixate ourselves on the weight of memories—dangerous or not. Such an idea does not question the significance of dangerous memories, but it certainly puts them in perspective. That is, we are reminded that all *dreamers of memory* dwell as though in a vacuum, as if there is no opposition to memory. However, in view of stories of subversion, the precarious opposition between past and present, remembering and forgetting, breaks down. In these terms, not to move around memories amounts to unwillingness to part company with concepts that deny human understanding and relationality (Arendt, 1978).

We are asked to consider, therefore, how teachers and students can create spaces that enable critical scenes of dangerous memories—not memories that freeze our relationality with others. In others words, teachers and students are called on to become critical witnesses of the others' suffering, that is, they need to learn to *see* things differently. Bearing witness, then, implies taking responsibility to become a transformative agent of awareness and reception of others' suffering (Oliver, 2001). Bearing witness is precisely what distinguishes the two stories narrated; presently, Yoni is willing to go as far as witnessing only the suffering of his people, while Anna struggles to witness the Other's suffering, despite the long socialization in believing that only her people have suffered. Anna is gradually becoming able to create new affective relations with *the enemy,* while Yoni is uncomfortable with the idea of witnessing *the enemy's* suffering and its relevance to terrorist acts. Becoming a witness can never be an individual achievement or an educational mission exclusively; it needs others to join, all working in the same direction. Becoming a witness needs to be supported by more than just the introduction of a dangerous memory; the work of the wider context that envelops the situation needs to be seen as supportive by the ones doing the work.

But this is not easily done. Yoni, for example, might be already ideologically positioned to allow for change and for giving up on his "I-only-victim" position—he is a conscientious objector already. Yet Yoni needs to conduct more than just one dialogue. Within the context of the group of teachers, Yoni sustains a dialogue with himself, one with his Jewish in-group (and its multiple positions), and another with the out-group (and the Palestinian multiple positions). He might fear them all, never knowing which argument will be interpreted, in what ways, by whom, with unpredictable consequences.

Consequently, the community is not always a vehicle for empathy and solidarity; in fact, it could just as easily be conceived as a construct of exclusion and dissensus. Community should not end up becoming a romantic delusion in contemporary educational and social discourse; hence, it needs to be acknowledged that communities do not include everyone. In other words, communities of memories (Simon & Eppert, 1997) or communities of resistance (Welch, 1985) are contested spaces, yet they can constitute spaces in which individuals may engage in *dialogues of witnessing*, that is, in discourses that formulate individuals' experiences away from prevailing oppressive forces and through collaborative efforts of interpretation and reinterpretation. The important point here is that unless students and teachers are engaged in a critical interrogation of collective memories, the risk is to fall again into the same cycle of foundational narratives as before. The goal is not symmetry in the acknowledgement of suffering—that is, *we* suffered; *they* suffered; we all suffered. Instead, the Other's narratives need to be witnessed, not absolved. Essentially, this is what Ibtisam asks Yoni to do with the knowledge Yoni acquires; it is also what Anna begins to explore when she becomes more receptive to the Other's story. To put it in a nutshell: Action that connects us with Others, in which a new We (that is not uncritically accepted) is always engaged in changing our taken-for-granted assumptions, stands in the sharpest possible opposition to the business of memory, which operates in egotistical terms. Based on our account here, anamnestic solidarity is an important part of the bridge that leads us from the latter to the former.

The two stories narrated here and the discussion about history, memory, and identity show not only the challenges but also the openings that may be created through witnessing the Other's suffering. Consequently, the pedagogical responsibility of educators is to create spaces in which students may explore collectively what it means to bear witness to the Other. Yet the challenges of doing so, as our stories suggest, highlight once again the need to account for the different sociohistorical contexts in the pedagogic work done around dangerous memories. Collective witnessing may lead—under some sociohistorical and educational conditions—to a political awareness and can attend to the work of transforming the future. However, collective witnessing is not a panacea and can easily lead to new forms of exclusion. Nevertheless, it is important to emphasize that collective witnessing is different from critical inquiry in that the latter often promotes educational individualism, while the former emphasizes the collective and the political, ethical, and critical aspects of collectivity (Boler & Zembylas, 2003). It is in this way that the classroom can become a potential place of political transformation, when students and teachers are acknowledged as witnesses who work through their affective connections to others, themselves, and their cowitnesses and find their grounding in anamnestic solidarity.

Not all individuals in a group hold similar positions; no individual is a societal dope; they might seem to be, but then what needs to be uncovered are the practices that get organized to allow for these commonalities to emerge. Zionist Israel has successfully developed such a machinery that has successfully produced the appearance of such dopes; the same success appears to be the case with the Greek-Orthodox Church and the nationalist political parties in both communities of Cyprus. Their *dopeness* should not be ridiculed and yet should be struggled against. Particularly in Israel, it is not easy to stand up against the official narratives in a country in which the Holocaust has been set as an ultimate justification—especially, when the young have not had direct experiences with it. The official powers are always present in the voices that surround (voices from the in-group) Yoni and Anna and successfully occupy a space that leaves little doubt even for the most courageous. A pedagogy of dangerous memories can be useful in the development of shared meanings created through intersubjective encounters. The focus of dangerous memories is not only on understanding the Other but also on developing critically shared meanings with the Other as the basis for relationship growth; this is precisely how a critical emotional praxis grounded in suffering is constituted.

Afterword: Beyond Trauma's Seduction

In this book, I have examined the politics of trauma in education by describing what trauma *does* in educational discourses and practices and by suggesting a rethinking of educational efforts so that trauma may be *translated* into critical emotional praxis. This rethinking constitutes an attempt with which educators and students can learn "to *mourn* what happened, bear witness to it, and yet move forward" (Kaplan, 2005, p. 147). It would be a mistake to view this attempt as a set of propositions; rather, what I have been concerned with throughout this book is to describe where the politics of trauma would leave us in education. In describing instances in which affective investments open up or bring closure to our connections with others, I have tried to show that various manifestations of affect—hatred, fear, shame, and anger—constitute ways of doing politics. Critical emotional praxis in schools, then, provides various kinds of *practical affective politics* (Thrift, 2005) in the sense of transforming existing pedagogical discourses and practices that often remain locked in historical categorizations.

In this afterword, I want to consider how the various pedagogical engagements proposed in the previous chapters constitute forms of *aporetic mourning* in educational theory and practice. In response to the dangers of nationalizing trauma and mourning both within schools and in the wider society, I emphasize that retrieving the politics of trauma in education and engaging in critical emotional praxis constitutes a *work of mourning* that contributes to strengthening our relationality to others; that is, the work of mourning as a pedagogical task makes possible new affective alliances. In fact, as I suggest, the constitution of a renewed affective community requires aporetic mourning, with no closure in sight, that comes only in relation to the Other. Learning to mourn loss (regardless of who suffers loss) in schools through engaging in critical emotional praxis can evoke public and school pedagogies that provide an alternative way of relating to otherness and trauma—not through remaining fixated on simply representing the other's or one's own trauma, but rather in the insistence on remaining *inconsolable* before loss. Therefore, I summarize the arguments put forward here by arguing that the recognition of our common vulnerability to loss, as the groundwork to the formulation of school and public pedagogies, can induce a powerful point of departure for political alignment with others; this alignment, I argue, has the potential to move us *beyond* trauma's seduction (Berlant, 2000).

THE WORK OF MOURNING

Mourning primarily provokes the question of what the lost other has been to us (Deutscher, 1998). Mourning, on Freud's account, is work and constitutes a psychic response to loss that reaches a conclusion—the *mourner* is able to work through grief, relinquish past attachments, and make new ones (Freud, 1917). Melancholia, by contrast, designates that mourning is an unfinished process in which the mourner has an open wound, incorporates the lost other into the self, and becomes perpetually attached to the absence of the lost one. For Freud, melancholia differs from mourning in that its work is never done; melancholia is, in a sense, *unsuccessful mourning*.

Although early theorizations have accepted uncritically Freud's distinction between successful and unsuccessful mourning, recent scholarship has reservations over establishing a universal, normal/ized pattern of mourning (Forter, 2004). Butler (1997b, 2004) and Derrida (1989, 2001), in particular, problematize the ethics and politics of mourning and expose the kind of critical interventions that are required to avoid the normalization of mourning. In Butler's work following September 11, there is a relentless critique of the hierarchies of public mourning and the repudiation of grief in the very discourse that seeks to redress it; Butler (2004) exposes the aggression "unleashed by national mourning in terms of the denial of vulnerability through the fantasy of national mastery" (Ziarek, 2006, p. 151). For Derrida, the work of mourning challenges us to acknowledge the alterity of the lost other without incorporating the other into the self or erasing the other; instead, the other is let to be in their absence and thus the work of mourning is without end. Derrida argues that mourning is not about incorporating the deceased, but their *gaze*. It is this gaze that makes us responsible before the deceased; a gaze, however, that makes mourning impossible because it always exceeds our capacity to make the deceased our own. Hence mourning is always related to the impossible incorporation of the other's gaze; an impossibility, nevertheless, that constitutes an infinite demand.

Despite their differences in theorizing the ethics and politics of mourning, both Butler and Derrida suggest that what binds us together is the experience of mourning, that is, the realization that we are all exposed and vulnerable, and are all engaged with bereavement (Engle, 2007). More importantly, I would argue that by relating critique to the work of mourning, Derrida and Butler call attention to the affective politics of loss and trauma as well as the need for critical intervention without either fetishizing narratives and practices of mourning or dismissing them. This politics of mourning does not work against the emotional ambivalence of trauma (chapters 3 and 7), but works within it and articulates a critical discourse from which to launch a renewal; that is, a restructuring of affective space in such a way that indeterminable mourning makes possible new affective connections.

Derrida and Butler on Mourning

As Zierek (2006) rightly observes, both Butler and Derrida foreground the impossibility of mourning but from two very different angles. Derrida describes the impossibility of mourning as an aporetic response to the loss of dear friends and the preservation of their alterity.[1] As Derrida (1996) writes: ". . . whoever thus works *at* the work of mourning learns the impossible—and that mourning is interminable. Inconsolable. Irreconcilable" (p. 172). He also invokes the concept of *impossible mourning*, in contrast to Freud's *normal* work of mourning, to emphasize that speaking *of* the other is impossible, because the other resists my knowledge and memory of him or her:

> we know our friend to be gone forever, irremediably absent, annulled to the point of knowing or receiving nothing himself of what takes place in memory . . . it would be unfaithful to delude oneself into believing that the other living *in us* is living in himself: because he lives *in us* and because we live this or that in his memory, in memory of him. (1989, p. 21, original emphasis)

Throughout his writings, Derrida evokes the possibility of an interiorization of what can never be interiorized, of what is always beyond us as a source of our responsibility. The gaze that is "in us" is never ours, yet it is the source of our infinite responsibility; this is the "unbearable paradox of fidelity" (2001, p. 159). We look at the dead, who have been reduced to the gaze "in us," and we are looked at by them, yet there is no symmetry between these gazes (Naas, 2003). Thus Derrida emphasizes that "this singular . . . affirmation [of mourning] *must* affirm the impossible . . . The impossible here is the other" (1989, p. 32). As he further questions how one can be faithful to the Other:

> What is an impossible mourning? . . . And as concerns the other in us . . . where is the most unjust betrayal? Is the most distressing, or even the most deadly infidelity that of a *possible mourning* which would interiorize within us the image, idol, or ideal of the other who is dead and lives only in us? Or is it that of the impossible mourning, which, leaving the other his alterity, respecting thus his infinite remove, either refuses to take or is incapable of taking the other within oneself, as in the tomb or the vault of some narcissism? (1989, p. 6, original emphasis)

In other words, for Derrida, any effort to "incorporate" the other "in us" is doomed to fail because we can never entirely free ourselves from the traces of the other, or if we do, we inevitably betray that other. On the one hand, a certain interiorization is unavoidable because the legacy of the dead is entrusted to memory; on the other hand, interiorizing the other leads to assimilation of the other's alterity. As Derrida (1995) puts it:

> I speak of mourning as the attempt, always doomed to fail (thus a constitutive failure, precisely) to incorporate, interiorize, introject, subjectivize the other in me. Even before the death of the other, the inscription in me of his or her mortality constitutes me. I mourn therefore I am, I am—dead with the death of the other, my relation is first of all plunged into mourning, a mourning that is moreover impossible. This is also what I call ex-appropriation, appropriation caught in a double bind; I must and I must not take the other into myself; mourning is an unfaithful fidelity if it succeeds in interiorizing the other ideally in me; that is, not respecting his or her infinite exteriority. (1995, p. 321).

For Derrida, then, mourning is faithful to the other when it is unfaithful in terms of interiorizing that other; a failure to mourn (that is, an unfaithful interiorization) is paradoxically faithful in that the alterity of the other is respected—the other is not assimilated. Derrida writes:

> We can only live this experience in the form of an aporia: the aporia of mourning . . . where the possible remains impossible. Where *success* fails. And where faithful interiorization bears the other and constitutes him in me . . . It makes the other a *part* of us. And then the other no longer quite seems to be the other because we grieve for him and bear him *in us* . . . And inversely, the *failure* succeeds: an aborted interiorization is at the same time a respect for the other as other, a sort of tender rejection, a movement of renunciation

> which leaves the other alone, outside, over there in his death, outside of us. (1989, p. 35, original emphasis)

It is from this perspective that Derrida argues, true mourning is impossible; if it succeeds, the otherness of the other is not respected. Perhaps to *truly* mourn, then, we must fail to mourn, be faithful to the indigestibility of the other. As Deutscher (1998) points out, Derrida converts the issue of mourning into one concerning the ethics of alterity. She further explains that Derrida's ethical affirmation of impossible mourning can be seen in the movement from an opposition between successful and failed mourning to its destabilization, so that all mourning may be seen as a generalized form of failed mourning. But in this form of failed mourning, we always harbor the other within us, because *we* are always already an interiorization of the other:

> If death comes to the other, and comes to us through the other, then the friend no longer exists except *in* us, *between* us. In himself, by himself, he is no more, nothing more. He lives only in us. But *we* are never *ourselves*, and between us, identical to us, a 'self' is never in itself or identical to itself. (1989, p. 28, original emphasis)

By contrast to Derrida's theorization of the impossibility of mourning, Butler exposes a very different impossibility, one that is intertwined with the psychic and social operations of power. According to Ziarek, the impossibility of mourning in Butler's work "is an effect of the disavowal of loss and the unconscious prohibition of grief" (2006, p. 144). The impossible manifests in the demand that losses (e.g., the loss of certain sexual attachments; the feeling of vulnerability after 9/11) are not to be grieved. Such disavowal of loss, Butler emphasizes, makes certain affective attachments ungrievable. Butler's concern, then, is to examine the circumstances in which psychic and social operations of power make mourning impossible. As Ziarek (2006) points out, in contrast to Derrida's ethical affirmation of impossible mourning, Butler's interest is in exposing the disavowal of an ethical affirmation.

In her work on *The Psychic Life of Power*, Butler (1997b) explores how mourning is rendered impossible as one of the main effects of the psychic and social operations of power. This tension is explored in the context of gender formation in homophobic culture in which, according to Butler, mourning is rendered impossible because the disavowal of loss (from the prohibition of homosexual attachments) is transformed into melancholic identifications. Butler builds on Freud's revision of his account on mourning according to which *all* mourning is at heart melancholic, because ambivalent incorporation becomes the only way under which objects can be *relinquished* (see Butler, 1997b, pp. 132–150). By incorporating the lost object through narcissistic identification with the lost Other, the melancholic ego seeks both to deny and to preserve its lost love (Forter, 2003). The incorporation of this ambivalent love/hate relation to the dead turns the violence of loss inward, thus "Melancholy is both the refusal of grief and the incorporation of loss, a miming of the death it cannot mourn" (1997b, p. 142).

Most importantly, however, the internalization of emotional ambivalence transforms the struggle into the conflict between the ego and the superego. As Butler (1997b) writes:

> When certain kinds of losses are compelled by a set of culturally prevalent prohibitions [e.g., homosexuality], we might expect a culturally prevalent form of melancholia. . . . And where there is no public recognition or discourse through which such a loss might be named and mourned, then melancholia takes on cultural dimensions of contemporary consequence. (p. 139)

This argument highlights the politically promising openings in Butler's account—especially in light of her appraisal of post–9/11 America—because her description emphasizes the psychosocial power relations under which the hegemony of melancholia flourishes. Butler analyzes the loss of the prohibited homosexuality and its transformation into a melancholic identification through exposing the dominant psychic mechanisms of compulsory heterosexuality. The loss of normative gender, then, according to Butler, is in fact prohibition against same-sex love in such a way that love cannot but be internalized in a normalized manner. Once this prohibition is publicly recognized, argues Butler, it becomes possible to work undoing these social norms as well as naming and mourning the losses. The goal of sexual politics is, then, minimizing melancholia through undoing social prohibitions.

Along these lines of thought, Butler (2004) in *Precarious Life* extends her analysis on mourning contextualizing it in post–9/11 America. One of Butler's main concerns in this book is the social prohibition of vulnerability through the fantasy of national mastery. Her point of departure is that "each of us is constituted politically in part by virtue of the social vulnerability of our bodies. . . . Loss and vulnerability seem to follow from our being socially constituted bodies, attached to others, at risk of losing those attachments, exposed to others, at risk of violence by virtue of that exposure" (p. 20). Butler argues that "we cannot . . . will away this vulnerability. We must attend to it" (p. 29). Writing about the U.S. shift to violence after 9/11, she poses the following questions:

> Is there something to be gained from grieving, from tarrying with grief, from remaining exposed to its unbearability and not endeavoring to seek resolution for grief through violence? . . . If we stay with the sense of loss, are we left feeling only passive and powerless, as some might fear? Or are we, rather, returned to a sense of human vulnerability, to our collective responsibility for the physical lives of one another? . . . To foreclose that vulnerability, to banish it, to make ourselves secure at the expense of every other human consideration is to eradicate one of the most important resources from which we must take our bearings and find our way. (p. 30)

To mourn, continues Butler, is to make mourning a resource for politics grounded on a consideration of the vulnerability of others. The denial of such vulnerability unleashes violence against others whereas its acknowledgment creates openings for an ethical encounter with others. "Then we might critically evaluate and oppose," she emphasizes, "the conditions under which certain human lives are more vulnerable than others, and thus certain human lives are more grievable than others" (p. 30). Once we consider how the psychosocial and political operations of power produce "who will be a grievable human" and what "acts" are "permissible" for "public grieving" (p. 37), then we may begin to realize how a prohibition of grieving others' lives extend the aims of violence. Examples of such prohibitions, says Butler, are how certain losses are not avowed as losses (e.g., the narratives of Arab lives killed elsewhere by brutal means). In this sense, we need to ask "about the conditions under which a grievable life is established and maintained, and through what logic of exclusion, what practice of effacement and denominalization" (p. 39).

Butler insists on a *common* vulnerability as the point of departure for a renewed politics of recognition that is fundamentally depended on existing norms of recognition. This implies that "to ask for recognition, or to offer it, is precisely not to ask for recognition for what one already is" (p. 44)—which is grounded on the historical categorizations of identity—but to solicit a transformation based on the relationality with the Other. Relationality forces us to appreciate, she argues, how far our ties with others constitute who we are. Hence Butler suggests that recognition of our own vulnerability to loss may indeed open up the potential for recognition of all humanity as vulnerable. The work of mourning helps us reconsider one's relation to the Other as a source of transformation. As she puts it, "one mourns when one accepts that by the loss one undergoes one will be changed, possibly forever"; it is a question, she says, of "submitting to transformation" (p. 21).

The juxtaposition of Derrida and Butler allows the intersection between the ethics and politics of mourning and provides a new insight into the relation between ethics and power: "one effect of the political regulation and prohibition of mourning would be a resolution of the aporia of an impossible ethical task" (Ziarek, 2006, p. 44). This resolution, maintains Ziarek, ranges from the repudiation of loss through the identification with the dead to the exclusion of ungrievable lives beyond the boundaries of the polis. To sustain the impossible as a paradoxical condition of public mourning, then public mourning has to remain aporetic: It too has to fail in order to succeed and it will succeed only if it risks failure. Only such discourse could prevent public mourning from being co-opted for the purposes of consolidating the fantasy of national mastery. The potentially radical innovation here, then, is the need to politicize psychoanalysis and analyze its intersections with power, suggesting that the work of mourning is precisely about the impossible incorporation of loss—neither its fetishization nor its dismissal.

Toward Aporetic Mourning in Education

The process of mourning is not without risks but, as Derrida (1997) makes clear, is "fragile and porous" (p. 13), and may end up collapsing under the weight of questions such as, "For whom does mourning take place? Who mourns, and for how long? Who, in effect, *decides*?" (Starling, 2002, p. 113). Butler also identifies the danger of personal narratives for unchecked narcissism and fetishism. She writes:

> In the U.S., we start the story by invoking a first-person narrative point of view, and tell what happened on September 11th. And it is that date, and the unexpected and fully terribly experience of violence that propels the narrative. . . . We have to shore up the first-person point of view, and preclude from the telling accounts that might involve a decentering of the narrative "I" within the international political domain. (2002, p. 58)

According to Butler, the first-person narrative entails the danger of *sentimentality*, which translates mourning into kitsch (Engle, 2007).

Following the novelist Milan Kundera's thinking, Engle writes that "Kitsch reflects a profound narcissism—a desire to insert oneself into a historical or otherwise significant moment" (p. 72). Kitsch, for example, enables the claim that a community's mourning (e.g., the U.S.'s mourning for 9/11) is essentially universal and that all mankind is moved by the same traumatic event. Out of this mass identification with and internalization of this traumatic event, however, borderlines between grieavable and ungrieavable lives are drawn, as Butler shows; *our* trauma is (and should be) grievable whereas others' traumas are not worthwhile to be grieved. Hence the danger is the erasure of difference, and eventually the nonrecognition of the Other in favor of homogeneity and absolute power.

Irrespective of the impact of these aberrant events on the larger political economy of victimization and aggression, the notion of *common suffering* helps facilitate the connection of losses between communities in conflict. This communal grieving is both a sociopolitical and a pedagogical act of mourning, both a private and public act of mourning, forging a connection in the affect of collective sorrow and vulnerability. In particular, the educational systems of conflicting communities have a lot to learn from juxtaposing different stories of loss and mourning, without, however, effacing the specificity of each loss or making losses equivalent.

In the last part of this afterword, I would like to take up the implications of aporetic mourning in relation to critical emotional praxis. Derrida's and Butler's accounts help us clarify the ethical and political orientations of mourning; their thoughts expose the conditions of possibility and impossibility through which mourning works. In relation to critical emotional praxis, a sustained endeavor to recognize not only one's own but also the Other's losses offers the potential to reinvent pedagogical spaces in which similarities and differences with others may be critically articulated and felt, toward constructing new shared imaginaries. For this purpose, I argue not for the elimination of aporias but in fact for an *aporetic mourning;* aporetic mourning nurtures a space that opens possibilities for reimagining the sense of community, identity, and trauma.

In other words, mourning has the potential to transcend individual empathic or vicarious suffering to produce new forms of *community* (Kaplan, 2005). Nancy (1991) has argued that "community is revealed in the death of others . . . because death itself is the true community of I's that are not egos. It is not a communion that fuses the egos into an Ego or a higher We. It is the community of others" (p. 15). Death binds us all together beyond any historical categories (race, gender, class); in the death of others, we "see" our own death. Therefore, witnessing the Other's suffering through mourning involves not just empathy and motivation to help, but the responsibility to recognize others through an awareness of injustice in the world. Inconsolable mourning for others' suffering, then, becomes a way of testifying to a fundamental commonality (Durrant, 2004). Whereas a historical account of trauma would turn it into a digestible narrative, allowing us to mourn and then move on, aporetic mourning resists this process of representation and forces us to confront the brute materiality of the suffering engendered by conflict: the irreconcilable sight of the dead body, stripped from any historical categorizations.

If mourning is an attempt, as Derrida has argued, to incorporate what cannot be incorporated, if mourning always negotiates between faithful and unfaithful fidelity to the Other, then mourning remains always aporetic and ambivalent and thus must always endure the aporias of knowing and not knowing. Therefore, honestly confronted in schools, aporetic mourning leads to neither fetishization nor dismissal of loss and trauma, but rather to an experience of ethical responsibility to the Other, that is, a form of critical emotional praxis. This solidarity with others lies at the heart of a pedagogy that forbids an unreflective and facile appropriation of others and demands critical encounter with the politics of trauma. The experience of such solidarity can lead educators and their students into a public discourse, in which conventional forms of collective identity and mourning are abandoned and replaced by new elements of identity that can anchor themselves in aporia. But how exactly can pedagogues formulate a learning to mourn in schools?

First, learning to mourn in schools requires that students be exposed to the individuality of each loss from the communities that are (or have been) involved in a conflict. Recognizing the Other's loss, argues Galuya (2006), opens up a space for connecting through an empathetic sorrow, despite not having the same loss. Galuya describes this as an "affective alliance with the other's sadness without claiming the other's specific loss as one's own" (p. 98). As Butler emphasizes, an inclusive mourning community has the potential to disrupt already formed national(ist) communities. Through each other's literature, media and the arts, students from communities in conflict can "come together over grief" and learn how to mourn "without effacing their specific grievances or eliding their particular loss" (Galuya, 2006, p. 98). They can also collect narratives of suffering from the other community, not only from their own community. These alliances are needed to change the conflicting ethos in schools and the wider society; it is a transformation of this ethos that might enable renewed coalitions, conviviality and solidarity between individuals and across communities.

Along these lines, several curriculum questions may be raised for exploration towards a learning to mourn in schools. For example: Under what circumstances can mourning take place? Who mourns, and for whom? In what ways can common suffering be a catalyst for reconciliation? What does it mean for people from conflicting communities to live together peacefully when they share a history of loss and suffering? What kind of selves do we need to reconcile with each other? It is in the context of these questions that curricula and pedagogies can embrace mourning as a process of learning to accept others as humans, without incorporating their narratives into predetermined categorizations.

However, considering the dangers discussed earlier, it is important not to assume that losses are interchangeable and mutually intelligible in an economy of exchange (Gregoriou, 2005). In such case, mutuality and interchangeability would eliminate the *aporia* that conditions the ethical encounter with the other. As Gregoriou writes:

> This mutual exchange of perspectives and transcendence of revenge is made possible by replicating the perspective of victimization. We suffered; they suffered; "we" all suffered. Once the autonomous peace of the suffering self is redeemed, once the confessional mourner negates the responsibility for listening to and joining the Other's mourning, the testimonial potential of mourning is neutralized. (p. 11)

The above aporia does not imply any symmetry in grief; common suffering does not mean the same suffering. In their search of openings for solidarity on the basis of common human suffering, schools and other social institutions and organizations can begin to study stories of trauma and mourning from each other's communities. It is in its practice that aporetic mourning can take the meaning of *critical hope* (Zembylas, 2007), while propelling individual and collective mourning into a process of forming new affective alliances.

Contrary to what might be expected, however, a pedagogy of learning to mourn in schools is by no means simply an alternative to nationalism or racism. Rather, its force and its risk derive precisely from within the structures it means to interrogate (Starling, 2002). What this entails is the risk of repeating the nationalism or rasicm it seeks to expose, yet it provides an opening to the otherness it excludes. And this opening comes precisely from discourses and practices that entail an ongoing struggle to encounter with the unknowable alterity of the Other (chapter 2). As Pensky (1989) argues, this opening happens when individuals and groups ground their relationality with others by telling the stories of those who suffered. The stories of suffering are shared, yet the process of mourning has no closure—not in the sense of paralyzing melancholia but in recognizing that common mourning has to have a beginning *now*, not after those in political positions move away from narratives that are locked in historical categorizations. The commitment and honest intent of individuals and groups from communities in conflict are needed to create social spaces for learning to mourn for/with others. Without undermining the power of hegemonic discourses on nationalism/racism/sexism and so on in public and educational circles, acts of common mourning signal the ethical responsibility of reorienting attention towards the (pedagogical) practices of relationality, conviviality, humility, and criticality.

The ideas discussed here explicate both the educational challenges and the openings created from an understanding of mourning as aporetic. Pedagogies that are mindful of the powers of violence, trauma, and mourning in the contemporary world require to help students become aware, both at an emotional and an intellectual level, of the shared meanings, visions, and ethical interdependence that can constrain as well as promote understanding and communal interaction. In each stage of this learning process, educators and their students are essentially asked to become witnesses "to that critical conjuncture when some of the selves who make up a culture loosen themselves from the security and comfort of old affiliations and identifications to make an unexpected 'gesture' of friendship toward all those on the other side of the fence" (Gandhi, 2006, p. 189). Aporetic mourning in the context of critical emotional praxis does not necessarily lay aside the orthodoxies of violence or hatred; however, aporetic mourning may enhance our capacities to form critical responses to others' traumas, and interrogate our persistent binary framing of trauma. Where aporetic mourning could take us, remains an open question; there is no finality in this task.

Notes

Introduction

1. An important clarification about the use of the term *other/Other* needs to be made here. The Other does not always signify the one who is marginalized. In general, I subscribe to Levinas's (1987b) idea that the Other is what I am not. For Levinas, the Other is in the first place the other human being who calls for our ethical responsibility. Throughout this book, I use one or the other (other/Other), often without distinguishing them; the distinctions are made carefully when it matters.
2. One caveat that concerns my understanding of emotions and affects needs to be made up front (see Zembylas, 2005, 2007). I view emotions not as internal or as qualities of the mind but as qualities of *action* (Lutz & Abu-Lughod, 1990). There is a rich tradition in the history of thinking about emotions-as-practices in disciplines like anthropology, sociology, communications, feminist studies, and cultural studies (Ahmed, 2004; Campbell, 1997; Lupton, 1998; Reddy, 2001). In adopting this account, emotions are viewed as performances that produce action within the context of particular social and political arrangements. By that it is meant that the words used to express emotions are not simply names for emotion entities, preexisting situations with coherent characteristics; rather, these words are themselves actions or ideological practices that serve specific purposes in the process of creating and negotiating reality (Lutz, 1988). The advantage of this position against others—for example, biological or cognitive theories of emotions—is in the acknowledgment that individuals are socially and culturally specific persons engaged in complex webs of power relations. Consequently, the degree to which emotion is socially valued and publicly expressed and the forms that such expressions are expected to take are historically variable (Reddy, 2001). Without an attention to emotions-as-practices in particular societies, any discussion about emotions constituting moral principles or providing indications about justice risks entailing an absolutist metaphysics. In particular, for the purposes of this book, the virtue of an account grounded in the notion of emotions-as-practices is that it makes available a conceptual scheme for conceiving traumatic or posttraumatic emotions as socially responsive. It is possible, therefore, to examine the emotional variations in response to traumatic events and experiences in relation to justice claims—a central issue of concern in this book (see chapters 1 and 6).

Simply and generally stated: If emotions matter in the everyday context, then they also matter for justice and the politics of trauma and loss.

I now need to reflect briefly on the distinction between emotions and affects. Savage (2004) asserts that emotions are simultaneously signs for bodily states and cultural categories. While it is cultural categories that enable individuals to put embodied sensations into words and make them into emotions, the embodied experiences that precede these cultural categories cannot be ignored (Lupton, 1998). Thus, it may be argued that *affects* are first experienced in the body and then named through social relations and culture (Leavitt, 1996; Massumi, 2002), while *emotions* cannot be reduced to biology, relations, or discourse alone but belong to all of these dimensions, as emotions are constituted through ongoing relational practices (Burkitt, 1997). In other words, emotions are in neither the individual nor the social but produce the very surfaces and boundaries that allow the individual and the social to be delineated as such (Ahmed, 2004).

Support for this approach on emotions and affects is found by applying feminist, critical, and poststructuralist theories to the study of emotion (e.g., see Campbell, 1997; Game & Metcalfe, 1996; Lupton, 1998; Lutz & Abu-Lughod, 1990; Massumi, 2002). These theories help us in the study of emotions and affects for two reasons. First, the theme of the *politics of emotions* is of major significance in the kind of thinking grounded in these theories. As Abu-Lughod and Lutz (1990) pointed out: "Emotional discourses are implicated in the play of power and the operation of a historically changing system of social hierarchy" (p. 15). Abu-Lughod and Lutz also noted that emotional discourses may establish, assert, challenge, or reinforce power and status differences. In other words, emotional discourses are historically constituted in the expression and enactment of hierarchies of power (Zembylas & Fendler, 2007). For example, the existing power hierarchies and the status quo of stereotypes for others (e.g., a nation is "our" eternal "enemy") are inextricably linked to hegemonic emotional discourses, that is, how emotions are constituted and reinforced in social and political life. Second, feminist, critical, and poststructuralist theories historicize the practices in which emotions are constituted. This perspective challenges an ahistorical conception of the subject as a site of emotional resistance or uncritical acceptance, a conception that blinds the subject to the ways power relations work to constitute feeling subjects. In other words, these theories analyze the transaction between larger sociopolitical forces and the internal psychic terrain of the individual, highlighting the ways this historicization can draw out changeable aspects of reality, thus allowing the possibility of ruptures and discontinuities of power differentials in the future (Boler, 1999; Zembylas, 2005, 2007). It is on this opening that several pedagogical interventions are grounded throughout this book.

3. Douglass and Vogler (2003) also use the term *witness studies* to emphasize the witnessing aspects of encounters with trauma. Trauma studies—

particularly Holocaust/Shoah related—written in the past two decades abound (e.g., see Caruth, 1996; Kaplan, 2005; LaCapra, 1994; Leys, 2000; Young, 1993).

4. Cyprus is a small country (9.251 sq km/3.572 sq miles) in the north-eastern part of the Mediterranean and its current population is around 1 million. The per capita income in 2006 was US$23,672 for the south (government-controlled area) and US$11,802 for the north (area administered by Turkish Cypriots). The political situation in Cyprus—known as the "Cyprus problem"—is dominated by the historical ethnic conflict between Greeks and Turks on the island. Cyprus emerged as an independent state in 1960 after a Greek-Cypriot guerrilla struggle against the British colonial rule. This anticolonial rebellion, however, did not aim at independence but at enosis, union with Greece. During that time, Turkish Cypriots, the largest minority on the island (18 percent), aimed at *taksim*, ethnic partition. The traumas of ethnic division, first in the 1960s, when the Turkish Cypriots were the main victims, and then in 1974, when Turkey invaded and divided the island, displacing many thousands of Greek Cypriots and Turkish Cypriots, came to signify an intractable conflict in the politics of the region. Since 1974, the Green Line has divided the island, separating the two communities—Greek Cypriots in the south and Turkish Cypriots in the north. The intractability of the conflict and of the persisting status quo, despite numerous diplomatic efforts, leads to many tensions on either side of the island. A major aspect of this tension comes from the fact that the two communities have lived apart for a long time, despite the partial lift of restrictions on movement across the Green Line in 2003. In April 2004, the Greek Cypriots rejected a proposed solution put in referendum by the United Nations, while the Turkish Cypriots accepted it; a few days later, Cyprus joined the European Union as a divided state with its problem still unresolved. Currently, the situation is marked by political stalemate, and various opinion polls in both communities indicate high pessimism for a solution in the near future.
5. The initial theorization of the notion *critical emotional praxis* began in the work I did with Megan Boler (Boler & Zembylas, 2003; Zembylas & Boler, 2002) and Sharon Chubbuck (Chubbuck & Zembylas, 2008). I would like to acknowledge their tremendous contributions in helping to formulate my thinking on some of the issues around this notion.

CHAPTER 2

1. As Kreuzer (2002) points out, the concept of *chosen trauma* differs somewhat from the concept of *collective trauma*. Despite the obvious difficulties one may have with the verb "chosen," (given that nobody chooses voluntarily to become a trauma victim), it does point to the fact that a group chooses to make a traumatic event the center of collective identification through a long-term socialization process (Svašek, 2005). On the other hand, collective trauma is

> a shattering, often violent event that affects a community of people . . . and that results from human behavior that is politically motivated and has political consequences. Such an event injures in one stab, penetrating all psychological defensive barriers of participants and observers, allowing no space for denial mechanisms and thus leaving those affected with an acute sense of vulnerability and fragility. Such traumatic events not only affect communities but sometimes also create communities, so that otherwise unconnected persons who share a traumatic experience seek one another out and develop a form of fellowship on the strength of that common tie. (Vertzberger, 1997, p. 864)

2. Some scholars—esepcially those whose work is grounded in the political strand (see next section)—have questioned the autonomous agency of collective memory and trauma, suggesting that collective memory and trauma may not survive psychologically from one generation to the next (Kansteiner, 2002, 2005; Winter, 2006).
3. It is certainly interesting to observe how some of Volkan's claims resonate with my own personal narrative in the introduction of this book.
4. As Verducci (2000) suggests, some theorists' understanding of *empathy* may be others' description of *sympathy*. An exploration of the historical terrain of empathy indicates substantive differences in the use and meaning of terms used.

CHAPTER 3

1. This action research project was conducted in the context of teaching Greek-Cypriot preservice teachers a foundational course on "Philosophy of Education." This course focused on concepts such as multiculturalism, diversity, peace, and reconciliation. The academic institution in which this project took place was located just a few kilometers from the Green Line separating the Greek Cypriots and the Turkish Cypriots. I decided to use action research as the vehicle to investigate my students' feelings of hatred because this methodology stressed the involvement of both the students and myself in creating collaboratively a more critical and meaningful educational experience. The cyclical nature of action research (Kemmis & McTaggart, 1988)—planning and action, observation and reflection—allowed us the opportunity to identify our concerns and goals, devise plans for action, evaluate the success of our efforts, and move forward to new concerns and questions. The data collection that seemed to fit more to this methodology and the recurring cycles of action research were: my own reflective journal as well as my students' journals; class notes and audio recordings; documents of student work; and official and unofficial interviews with my students on various aspects related to their feelings about the Other (i.e., Turkish Cypriots).
2. The conflict in Cyprus is an example of interethnic and intercultural

dissonance, one with a long history that goes back several centuries. Looking at the history of Cyprus, one can easily find competing nationalist discourses of how each community represents the Other (see also chapters 6 and 8). The common themes of these narratives focus on the unimaginable violence and historical trauma that one side has inflicted on the other, and emphasize that it is imperative to remember the suffering of the past. One example of these different narratives is that Greek Cypriots talk about their suffering as a result of the "1974 Turkish invasion," while Turkish Cypriots welcome Turkey's 1974 intervention as a "happy peace operation" (Mutlu Bariş Harekâti) that saved Turkish Cypriots from the evil hands of Greek Cypriots. After the war of 1974, the Greek Cypriots have invested a lot in creating a strong visual/educational culture of not forgetting their "Turkish occupied territories," a campaign that has become known as "I do not forget and I struggle" (Den Xechno kai Agonizomai). Similarly, Turkish Cypriots have established museums such as the Museum of Barbarism (Barbarlik Musesi) in which they portray the acts of barbarism that they suffered from Greek Cypriots before the 1974 Turkish intervention and particularly the Events of the Bloody Christmas of 1963 (Kanli Noel). Turkish Cypriots have a similar slogan concerning the importance of not forgetting the past, "We don't forget" (unutmayacağiz), which refers to the bloody events of 1963. Over the years, each community has formed negative opinions of the Other not only through the official education system but also through social imaginary, the mass media, and familial and social circles.

CHAPTER 4

1. Cyprus joined the European Union in 2004; the arrival of asylum seekers and immigrants from the Middle East and Asia has increased since Cyprus' accession to the European Union, as Cyprus has become the most southern border of Europe that is also a gateway to the Middle East and Asia.
2. Similarly, Parkin (1986) refers to what he calls "raw fear," which has a more free-floating and unpredictable character. Bourke (2006) also claims that this more nebulous state of fear is due to the decline of tangible threats to corporeal existence that were brought about by war in earlier eras.
3. As I am writing this manuscript, the news in Cyprus is dominated by both government and unofficial voices that the country is at serious risk of being overrun by immigrants. For example, a newly formed organization, the Movement for the Salvation of Cyprus, is concerned that the growing number of immigrants on the island will irrevocably change the island's demographics over the next 30 years. The narrative that is being utilized by this organization is familiar: "We are fighting for our national identity. . . . We are sounding the alarm bell and saying 'wake up, you're losing your country'" said one of the organizers in a gathering of this

Movement. (This quote is reported in a newspaper article titled "Wake up, you're losing your country" by Alexia Saoulli, *Cyprus Mail,* December 12, 2007, accessed at http://www.cyprus-mail.com/news/). This narrative works through *othering;* the immigrants are those who are "not us" and thus endanger what is ours. Thus, they call all of "you," as the legitimate citizens of this country, to adopt this "you" against "them" and defend the country from these others. As I have already pointed out in this and in earlier chapters, this narrative invites those who belong to "us" through working on emotions, and particularly fear. As one of the individuals present in this gathering said: "We have been terrorized but this is the truth. These are the facts." These brief examples evoke a we that is only available to the Greek Cypriots and equates the vulnerability of Cyprus with the vulnerability of the Greek Cypriots: "Wake up, you're losing your country." Similar examples are found in other European countries (e.g., see Ahmed, 2004).

Chapter 5

1. My discussion here implies that shame is necessary for empathy. Undoubtedly, the conditions of possibility for engaging with shame in schools must be constantly interrogated, not only because recognizing shame may help us empathize with others but also because of *empty empathy* (Kaplan, 2005). Hearing trauma narratives in schools that go beyond the dichotomy pride/shame is a challenging task. Therefore, it may be argued that under some conditions shame may help us empathize with the Other, but there is no assurance that sentimentalism will not compromise struggles to engage in a critical self-evaluation of shameful acts. (Some of these conditions are teased out in chapter 1.)
2. The Australian government's formal apology for the past wrongs inflicted on the indigenous Aboriginal population (on February 13, 2008) is certainly a historic and moving event. It may be argued that the expression of apology for the "indignity and degradation," and "the profound grief, suffering and loss" inflicted on the country's Aboriginal population (see the Prime Minister's speech at http://www.smh.com.au/news/national/kevin-rudds-sorry-speech/2008/02/13/1202760379056.html) constitutes a form of regret for past wrongdoings. Yet it can also be argued that saying sorry does not subvert the marginalization of and the social injustices against the Aboriginal population. Nevertheless, this apology is a historic step in recognizing past wrongdoings, thus it remains to be seen whether saying sorry has any real, material implications for the reconciliation process in Australia.

Chapter 6

1. Violence, in its broadest sense, includes physical, psychological, and structural violence and can be caused by thoughts, words, and deeds

(Harris, 1996). More recently, Francis (2006) distinguishes between "direct violence," which is physical, emotional, and psychological; "structural violence," which refers to deliberate politics and structures that cause suffering and harm; and "cultural violence," which refers to cultural norms and practices that create discrimination, injustice, and suffering.

2. The state or concept of peace is certainly not monolithic; in peace theory, the absence of war (*negative peace*) is distinguished from harmony and social justice (*positive peace*).
3. Berlowitz (1994) argued that there are three major ways to promote peace education in schools: peacekeeping (peace through strength); peacemaking (peace through dialogue); and peace building (peace through creating conditions necessary for peace such as developing attitudes and dispositions for nonviolent communication). Peace through strength (peacekeeping) relies on force and threats of force to deter violence or punish aggressors; peace through dialogue (peacemaking) teaches skills to resolve conflicts through communication; and finally, peace building relies on building the conditions for creating a culture of peace and thus prevents conflicts from becoming violent (Galtung & Jacobsen, 2000). Traditionally, schools have relied upon a peacekeeping response to the problems of violence, that is, the promotion of peace through control of student behavior (Harris, 1999, 2002b). Harris criticizes this approach because it blames individuals for the dysfunctional behaviors they have adopted from their social environment, instead of looking at the problem of violence at a wider social and political level. Peacemaking requires dialogue and is usually accomplished through teaching conflict resolution skills and implementing peer mediation programs that help resolve conflict in a win-win manner. Again, this approach can be criticized on the same grounds as the previous one; it sees the problems of violence as residing in individuals and hopes to solve violence by learning how to avoid violent behaviors. Finally, peace building teaches students how to live peacefully by creating the conditions for peace such as attitudes, habits/behaviors, and dispositions for positive peace. The goal, then, is not just to stop violence but also to empower young people to engage in nonviolent communication (Harris, 2002b).
4. *Moral exclusion theory* (Opotow, 1990, 1995, 2001) refers to the limited applicability of justice underlying conflicts and difficult social problems. This theory makes an explicit connection between peace building and social justice in its emphasis on fairness, resource sharing, and concern for the well-being of all. Those who are morally excluded from the scope of justice are deprived and exploited as expendable and undeserving of fairness and community resources that could foster their well-being. Opotow et al. (2005) argue that peace education should be used as a tool for moral inclusion and for bringing about a world in which justice applies to all.

CHAPTER 8

1. Another striking gap in Halbwachs's framework was the absence of analysis of collective trauma. As Paez, Basabe, and Gonzalez (1997) have pointed out, "Halbwachs's framework does not propose clear mechanisms that allow for an explanation of how societies remember conflict-ridden collective events with negative aspects and conflicting meanings" (p. 162). This omission, writes Becker (2005), is all the more noteworthy given that,

   > At the very time when an arsenal of unprecedented memories born of the consequences of the Great War was being put into place, here was a man theorizing the notion of collective memory while simultaneously forgetting, in the numerous examples punctuating his work on the subject, to think about the weight of the recent past including his own personal past. (p. 103)

   Paradoxically, Halbwachs seems to have repressed his own trauma (he had seen the battlefields and corpses of the Great War) so that it led him to exclude the phenomenon of trauma entirely from his treatise on memory (Kosicki, 2007).
2. An interesting issue here is the question of "dangerous for whom?" Dangerous memories, as we argue later in the chapter, are potentially reconciliatory, therefore, one may ask how much they are really "dangerous." They are indeed threatening to official memory and vested national interests, and temporarily threatening to the experiencer's conventional understanding. We could, therefore, have used the term "disruptive" rather than "dangerous." However, we want to retain the term "dangerous" for strategic purposes and so we adopt Ostovich's (2002, 2005) use of it, because we believe it is important to raise attention (politically speaking) to the fact that reconciliatory memories are threatening to hegemonic discourses. (See next section for further analysis of this idea.)
3. Given that the social and political context about Cyprus has been presented in previous chapters, a similar discussion is offered here about Israel and the Palestinian-Israeli conflict. The state of Israel has institutionalized itself through the establishment of public education, the standardization of law, and the development of a secular equivalent to church (Ben-Amos & Bet-El, 1999; Gellner, 1997; Handelman, 1990). Like many other modern nation-states, it is the product of an invented tradition (Hobsbawm, 1983). The Jewish-Palestinian conflict remains the most potentially explosive of conflicts in Israel, placing the Jewish majority (80 percent of the population) and the Palestinian (primarily Moslem) minority (20 percent) in a situation plagued with tragedy and suffering. For the most part, Israel as an ethnic democracy (Smooha, 1996) has not welcomed the active participation in political, cultural, or social spheres of anyone other than its legitimate invented community (Anderson, 1991) of Jews. Israeli Palestinians, though officially offered full rights as

citizens, have chronically suffered as a putatively hostile minority, with little political representation and a debilitated social, economic, and educational infrastructure (Ghanem, 1998). The rather optimistic political outlook in the wake of the Oslo agreements that characterized the period during which the bilingual schools were established has changed radically since October 2000 with a renewed wave of violent conflict that has taken the lives of many civilians, both in Israel and in the Palestinian Authority. Even after the implementation of the recent disengagement policy in Gaza, no one really knows what the future holds for the area. Though riddled with conflict and social cleavages, Israel must attempt to meet the often-competing requirements of a multiethnic national-religious society. The sociopolitical conflicts are reflected in the Israeli educational system, which is divided into separate educational sectors: Non-religious Jewish, Religious National Jewish, Orthodox Jewish, and Arab, all under the umbrella of the Israeli Ministry of Education (Sprinzak, Segev, Bar, & Levi-Mazloum, 2001).

4. Those interested might want to review some of the published works based on these data: see Bekerman (2003, 2004, 2005 a, 2005 b, 2007b) and Zembylas (2006b, 2007, Zembylas & Karahasan, 2006).
5. However, it needs to be made clear that Greek-Cypriot teachers are not presently encouraged to arrange meetings between Greek-Cypriot and Turkish-Cypriot schoolchildren, mainly for political reasons (i.e., fear of recognizing the status of a "state" in the other side), and thus any encounters are either accidental or take place in the realm of individual initiatives.

AFTERWORD

1. Although Derrida's analysis is essentially based on what might be called private mourning, individual mourning, in this case his mourning of close friends and colleagues, there are points raised that can relate these reflections to public mourning (Naas, 2003).

References

Abu-Lughod, L., & Lutz, C. A. (1990). Introduction: Emotion, discourse, and the politics of everyday life. In C. Lutz & L. Abu-Lughod (Eds.), *Language and the politics of emotion* (pp. 1–23). Cambridge, UK: Cambridge University Press.

Ahmed, S. (2000). *Strange encounters: Embodied others in post-coloniality.* London: Routledge.

Ahmed, S. (2001). The organization of hate. *Law and Critique, 12,* 345–365.

Ahmed, S. (2003). The politics of fear in the making of worlds. *Qualitative Studies in Education, 16,* 377–398.

Ahmed, S. (2004). *The cultural politics of emotion.* Edinburgh, UK: Edinburgh University Press.

Ahmed, S. (2005). The skin of the community: Affect and boundary formation. In T. Chanter & E. Plonowska Zierek (Eds.), *Revolt, affect, collectivity: The unstable boundaries of Kristeva's polis* (pp. 95–111). Albany: State University of New York Press.

Aho, J. A. (1994). *This thing of darkness: A sociology of the enemy.* Seattle: University of Washington Press.

AKTI-Project and Research Centre. (2004). *Report on the history and literature textbooks of 6th grade in terms of promoting violence and nationalism.* Nicosia, Cyprus: AKTI.

Albrecht-Crane, C. (2005). Pedagogy as friendship: Identity and affect in the conservative classroom. *Cultural Studies, 19,* 491–514.

Allport, G. W. (1954). *The nature of prejudice.* Cambridge, MA: Addison-Wesley.

Altheide, D. (2002). *Creating fear: News and the construction of crisis.* New York: Aldine De Gruyter.

Altheide, D. (2006). Terrorism and the politics of fear. *Cultural Studies/ Critical Methodologies, 6,* 415–439.

Anderson, B. (1991). *Imagined communities: Reflections on the origins and spread of nationalism.* London: Verso.

Appiah, K. A. (1994). Identity, authenticity, survival: Multicultural societies and social reproduction. In A. Gutmann (Ed.), *Multiculturalism* (pp. 149–163). Princeton, NJ: Princeton University Press.

Appiah, K. A., & Gutmann, A. (1996). *Color conscious: The political morality of race.* Princeton, NJ: Princeton University Press.

Apple, M. (1979). *Ideology and curriculum.* London: Routledge.
Apprey, M. (1999). Reinventing the self in the face of received transgenerational hatred in the African American community. *Journal of Applied Psychoanalytic Studies, 1,* 131–143.
Arendt, H. (1978). *The life of the mind.* New York: Harcourt Brace & Company.
Aronson, E. (1992). *The social animal.* New York: W. H. Freeman and Company.
Ashley, M. J. (1998). *Interruptions: Mysticism, politics, and theology in the work of Johann Baptist Metz.* Notre Dame, IN: University of Notre Dame Press.
Assmann, A. (2006). History, memory and the genre of testimony. *Poetics Today, 27,* 261–273.
Attalides, M. (1979). *Cyprus: Nationalism and international politics.* Edinburgh, UK: Q Press.
Babad, E. (1993). Pygmalion—25 years after: Interpersonal expectations in the classroom. In P. D. Blanck (Ed.), *Interpersonal expectations: Theory, research, and applications* (pp. 125–153). London: Cambridge University Press.
Bakhtin, M. M. (1981). *The dialogic imagination: Four essays by M. M. Bakhtin* (C. Emerson & M. Holquist, Trans.). Austin: University of Texas Press.
Bakhtin, M. M. (1984). *Problems of Dostoevsky's poetics* (C. Emerson, Trans.). Minneapolis: University of Minnesota Press.
Bakhurst, D. (1995). Wittgenstein and social being. In D. Bakhurst & C. Sypnowich (Eds.), *The social self* (pp. 30–46). London: Sage.
Balibar, E. (2003). *We, the people of Europe? Reflections on transnational citizenship* (J. Swenson, Trans.). Princeton, NJ: Princeton University Press.
Barbalet, J. (1998). *Emotion, social theory and social structure.* Cambridge, UK: Cambridge University Press.
Barkan, E. (2000). *The guilt of nations: Restitution and negotiating historical injustices.* New York: Norton.
Bar-Tal, D. (2000). From intractable conflict through conflict resolution to reconciliation: Psychological analysis. *Political Psychology, 21,* 351–365.
Bar-Tal, D. (2003). Collective memory of physical violence: Its contribution to the culture of violence. In E. Cairns & M. D. Roe (Eds.), *The role of memory in ethnic conflict* (pp. 77–93). Houndmills, UK: Palgrave Macmillan.
Bartky, S. (1990). *Femininity and domination.* New York: Routledge.
Bauman, Z. (1991). *Modernity and ambivalence.* Cambridge, UK: Polity Press.
Bauman, Z. (1992). *Intimations of postmodernity.* London: Routledge.
Bauman, Z. (2006). *Liquid fear.* Cambridge, UK: Polity Press.
Baures, M. M. (1996). Letting go of bitterness and hate. *Journal of Humanistic Psychology, 36,* 75–90.
Becker, A. (2005). Memory gaps: Maurice Halbwachs, memory and the Great War. *Journal of European Studies, 35,* 102–113.

Bekerman, Z. (2003). *The influence of teachers' multicultural and bilingual perspectives on integrated Palestinian-Jewish education.* Paper presented at the Teaching and Learning for Intercultural Understanding, Human Rights and a Culture of Peace, Jyvaskyla, Finland.

Bekerman, Z. (2004). Multicultural approaches and options in conflict ridden areas: Bilingual Palestinian-Jewish education in Israel. *Teachers College Record, 106,* 574–610.

Bekerman, Z. (2005a). Are there children to educate for peace in conflict-ridden areas? A critical essay on peace and coexistence education. *Intercultural Education, 16,* 235–246.

Bekerman, Z. (2005b). Complex contexts and ideologies: Bilingual education in conflict-ridden areas. *Journal of Language Identity and Education, 4,* 1–20.

Bekerman, Z. (2007a). Collective memory and obstacles to reconciliation efforts in Israel. In S. T. Jassal & E. Ben-Ari (Eds.), *The partition motif in contemporary conflicts* (pp. 323–343). London: Sage.

Bekerman, Z. (2007b). Rethinking intergroup encounters: Rescuing praxis from theory, activity from education, and peace/co-existence from identity and culture. *Journal of Peace Education, 4,* 21–37.

Bekerman, Z., & Maoz, I. (2005). Troubles with identity: Obstacles to coexistence education in conflict ridden societies. *Identity: An International Journal of Theory and Research, 5,* 341–357.

Bekerman, Z., & McGlynn, C. (Eds.). (2007). *Addressing ethnic conflict through peace education.* New York: Palgrave Macmillan.

Bekerman, Z., & Tatar, M. (2005). Overcoming modern-postmodern dichotomies: Some possible benefits for the counseling profession. *British Journal of Guidance and Counseling, 33,* 411–421.

Ben-Amos, A., & Bet-El, I. (1999). Holocaust day and Memorial day in Israeli schools: Ceremonies, education and history. *Israel Studies, 4,* 258–284.

Benjamin, S. (2002). *The micropolitics of inclusive education.* Buckingham, UK: Open University Press.

Benjamin, W. (1968). *Illuminations* (H. Zohn, Trans.). New York: Harcourt, Brace & World.

Bennett, J. (2001). *The enchantment of modern life: Attachments, crossings, and ethics.* Princeton, NJ: Princeton University Press.

Bennett, J. (2005). *Empathic vision: Affect, trauma, and contemporary art.* Stanford, CA: Stanford University Press.

Berlak, A. (1999). Teaching and testimony: Witnessing and bearing witness to racisms in culturally diverse classrooms. *Curriculum Inquiry, 29,* 99–127.

Berlak, A. (2004). Confrontation and pedagogy: Cultural secrets and emotion in antioppressive pedagogies. In M. Boler (Ed.), *Democratic dialogue in education: Troubling speech, disturbing silence* (pp. 123–144). New York: Peter Lang.

Berlant, L. (2000). The subject of true feeling: Pain, privacy, and politics. In S. Ahmed, J. Kilby, C. Lury, M. McNeil, & B. Skeggs (Eds.), *Transformations: Thinking through feminism* (pp. 33–47). London: Routledge.

Berlant, L. (2001). Trauma and ineloquence. *Cultural Values, 5*(1), 41–58.

Berlant, L. (Ed.). (2004). *Compassion: The culture and politics of an emotion.* New York: Routledge.

Berlowitz, M. (1994). Urban educational reform: Focusing on peace education. *Education and Urban Society, 27*(1), 82–95.

Best, S., & Kellner, D. (1991). *Postmodern theory: Critical interrogations.* New York: Guilford Press.

Bhabha, H. (1990). Dissemination: Time, narrative, and the margins of the modern nation. In H. Bhabha (Ed.), *Nation and narration* (pp. 291–322). London: Routledge.

Bhabha, H. (1994). *The location of culture.* London: Routledge.

Bingham, C. (2006). Before recognition and after: The educational critique. *Educational Theory, 56,* 325–344.

Biton, Y., & Salomon, G. (2006). Peace in the eyes of Israeli and Palestinian youths: Effects of collective narratives and peace education program. *Journal of Peace Research, 43,* 167–180.

Blee, K. (2002). *Inside organized racism: Women in the hate movement.* Berkeley: University of California Press.

Blee, K. (2004). Positioning hate. *Journal of Hate Studies, 3,* 95–105.

Bodnar, J. (1992). *Remaking America: Public memory, commemoration, and patriotism in the twentieth century.* Princeton, NJ: Princeton University Press.

Boler, M. (1999). *Feeling power: Emotions and education.* New York: Routledge.

Boler, M. (2001). *LoveOnLine: Educating eros in digital education.* Paper presented at the annual meeting of the American Educational Research Association, Seattle, WA.

Boler, M. (Ed.). (2004). *Democratic dialogue in the classroom: Troubling speech, disturbing silence.* New York: Peter Lang.

Boler, M., & Zembylas, M. (2003). Discomforting truths: The emotional terrain of understanding differences. In P. Tryfonas (Ed.), *Pedagogies of difference: Rethinking education for social justice* (pp. 110–136). New York: Routledge.

Boltanski, L. (1999). *Distant suffering: Morality, media and politics.* Cambridge, UK: Cambridge University Press.

Booth, W. J. (1999). Communities of memory: On identity, memory and debt. *American Political Science Review, 93,* 249–263.

Bouchard, N. (2002). A narrative approach to moral experience using dramatic play and writing. *Journal of Moral Education, 31,* 407–422.

Bourdieu, P. (2000). *Pascalian meditations* (R. Nice, Trans.). Stanford, CA: Stanford University Press.

Bourguignon, E. (2005). Memory in an amnesic world: Holocaust, exile, and the return of the suppressed. *Anthropological Quarterly, 78,* 63–88.

Bourke, J. (2006). *Fear: A cultural history.* Emeryville, CA: Shoemaker Hoard.

Boyarin, D., & Boyarin, J. (1995). Diaspora: Generation and the ground of Jewish identity. In K. A. Appiah & H. L. Gates (Eds.), *Identities* (pp. 305–337). Chicago: University of Chicago Press.
Brison, S. (1999). Trauma narratives and the remaking of the self. In M. Bal, J. Crewe, & L. Spitzer (Eds.), *Acts of memory: Cultural recall in the present* (pp. 37–54). Hanover, NH: University Press of New England.
Britzman, D. (1998). *Lost subjects, contested objects: Toward a psychoanalytic inquiry of learning.* New York: State University of New York Press.
Broome, B. J. (1991). Building shared meaning: Implications of a relational approach to empathy for teaching intercultural communication. *Communication Education, 40,* 235–249.
Broome, B. J. (1993). Managing differences in conflict resolution: The role of relational empathy. In D. J. Sandole & H. van der Merwe (Eds.), *Conflict resolution theory and practice: Integration and application* (pp. 97–111). New York: Manchester University Press.
Brown, W. (1995). *States of injury: Power and freedom in later modernity.* Princeton, NJ: Princeton University Press.
Bryant, R. (1998). An education in honor: Patriotism and the schools of Cyprus. In V. Calotychos (Ed.), *Cyprus and its people: Nation, identity, and experience in an unimaginable community, 1955–1997* (pp. 69–84). Boulder, CO: Westview Press.
Bryant, R. (2001). An aesthetics of self: Moral remaking and Cypriot education. *Comparative Study of Society and History, 43,* 583–614.
Bryant, R. (2004). *Imagining the modern: The cultures of nationalism in Cyprus.* London: I. B. Tauris.
Buonfino, A. (2004). Between unity and plurality: The politicization and securitization of the discourse of immigration in Europe. *New Political Science, 26,* 23–49.
Burbules, N. C. (2000). The limits of dialogue as critical pedagogy. In P. P. Trifonas (Ed.), *Revolutionary pedagogies: Cultural politics, instituting education and the discourse of theory* (pp. 251–273). New York: Routledge-Farmer.
Burke, K. (1969). *A grammar of motives.* Berkeley: University of California Press.
Burke, P. (1989). History as social memory. In T. Butler (Ed.), *History, culture, and the mind* (pp. 97–114). New York: Basil Blackwell.
Burkitt, I. (1997). Social relations and emotions. *Sociology, 31,* 37–56.
Burns, B., Busby, C., & Sawchuk, K. (Eds.). (1999). *When pain strikes.* Minneapolis: University of Minnesota Press.
Burns, R. J., & Aspeslagh, R. (1996). *Three decades of peace education around the world: An anthology.* New York: Garland.
Burrow, S. (2005). The political structure of emotion: From dismissal to dialogue. *Hypatia, 20*(1), 27–43.
Butler, J. (1997a). *Excitable speech: A politics of the performative.* London: Routledge.
Butler, J. (1997b). *The psychic life of power: Theories in subjection.* Stanford, CA: Stanford University Press.

Butler, J. (2002). Explanation or exoneration, or what we can hear. *Grey Room, 7,* 177–188.

Butler, J. (2004). *Precarious life: The powers of mourning and violence.* London: Verso.

Cairns, E., & Hewstone, M. (2002). The impact of peacemaking in Northern Ireland on intergroup behaviour. In G. Salomon & B. Nevo (Eds.), *Peace education: The concepts, principles, and practices around the world* (pp. 217–228). Mahwah, NJ: Lawrence Erlbaum Associates.

Campbell, S. (1994). Being dismissed: The politics of emotional expression. *Hypatia, 9*(3), 46–65.

Campbell, S. (1997). *Interpreting the personal: Expression and the formation of feelings.* Ithaca, NY: Cornell University Press.

Card, C. (2002). *The atrocity paradigm: A theory of evil.* New York: Oxford University Press.

Caruth, C. (Ed.). (1995). *Trauma: Explorations in memory.* Baltimore: Johns Hopkins University Press.

Caruth, C. (1996). *Unclaimed experience: Trauma, narrative, and history.* Baltimore: Johns Hopkins University Press.

Cherry, S. (2004). Forgiveness and reconciliation in South Africa. In F. Watts & L. Guil (Eds.), *Forgiveness in context: Theology and psychology in creative dialogue* (pp. 160–177). London: Continuum.

Chizhik, E. W., & Chizhik, A. W. (2002). A path to social change: Examining students' responsibility, opportunity, and emotion toward social justice. *Education and Urban Society, 34,* 283–297.

Chubbuck, S., & Zembylas, M. (2008). The emotional ambivalence of socially just teaching: A case study of a novice urban school teacher. *American Educational Research Journal, 45(2),* 274–318.

Cohen, C. E. (1997). *A poetics of reconciliation: The aesthetic mediation of conflict.* Unpublished doctoral dissertation, University of New Hampshire.

Cole, E. (Ed.). (2007). *Teaching the violent past: History education and reconciliation.* Lanham, MD: Rowman & Littlefield.

Confino, A. (1997). Collective memory and cultural history: Problems of method. *American Historical Review, 102,* 1386–1403.

Constantinou, C. M. (2001). The Cypriot state(s) *in situ*: Cross-ethnic contact and the discourse of recognition. *Global Society, 15,* 125–148.

Conway, B. (2003). Active remembering, selective forgetting, and collective identity: The case of Bloody Sunday. *Identity: An International Journal of Theory and Research, 3,* 305–323.

Craib, I. (1994). *The importance of disappointment.* London: Routledge.

Danesh, H. B. (2006). Towards an integrative theory of peace education. *Journal of Peace Education, 3,* 55–78.

Davies, L. (2004). *Education and conflict: Complexity and chaos.* London: RoutledgeFalmer.

Davies, W. (2000). *Overcoming anger and irritability: A self-help guide using cognitive behavioral techniques.* New York: New York University Press.

Deleuze, G. (1964/1972). *Proust and sings* (R. Howard, Trans.). New York: George Braziller.

Denzin, N., & Lincoln, Y. (Eds.). (2003). *9/11 in American culture*. Latham, MD: Rowman & Littlefield.

de Rivera, J., Kurrien, R., & Olsen, N. (2007). The emotional climate of nations and their culture of peace. *Journal of Social Issues, 63,* 255–271.

de Rivera, J., & Páez, S. (2007). Emotional climate, human security, and cultures of peace. *Journal of Social Issues, 63,* 233–253.

Derrida, J. (1978). *Writing and difference*. London: Routledge & Kegan Paul.

Derrida, J. (1989). *Mémoires: For Paul de Man* (C. Lindsay, E. Cadava, J. Culler, & P. Kamuf, Trans.). New York: Columbia University Press.

Derrida, J. (1995). *Points . . . Interviews, 1974–1994* (E. Weber, Ed.). Stanford, CA: Stanford University Press.

Derrida, J. (1996). By force of mourning. *Critical Inquiry, 22,* 171–192.

Derrida, J. (1997). *The politics of friendship* (G. Collins, Trans.). London: Verso.

Derrida, J. (2001). *The work of mourning*. Chicago: University of Chicago Press.

Deutscher, P. (1998). Mourning the other, cultural cannibalism, and the politics of friendship (Jacques Derrida and Luce Irigaray). *Differences: A Journal of Feminist Cultural Sciences, 10,* 159–186.

Dhillon, P. (2000). The sublime face of just education. In P. Dhillon and P. Standish (Eds.), *Lyotard: Just education* (pp. 110–124). London: Routledge.

Donald, J. (1992). *Sentimental education: Schooling, popular culture and the regulation of liberty*. London: Verso.

Douglass, A., & Vogler, T. (2003). *Witness and memory: The discourse of trauma*. New York: Routledge.

Doxtader, E., & Villa-Vicencio, C. (Eds.). (2003). *Through fire with water: The roots of division and the potential of reconciliation in Africa*. Cape Town, South Africa: David Philip.

Duijzings, G. (1999). *Religion and the politics of identity in Kosovo*. London: Hurst and Company.

Durrant, S. (2004). *Postcolonial narrative and the work of mourning: J. M. Coetzee, Wilson Harris and Toni Morrison*. Albany: State University of New York Press.

Duvenage, P. (1999). The politics of memory and forgetting after Auschwitz and apartheid. *Philosophy and Social Criticism, 25*(3), 1–28.

Dyson, M. E. (1994). Essentialism and the complexities of racial identity. In D. T. Goldberg (Ed.), *Multiculturalism: A critical reader* (pp. 218–229). New York: Blackwell.

Eisenstein, Z. (1996). *Hatreds: Racialized and sexualized conflicts in the 21st century*. New York: Routledge.

Ekman, P., & Davidson, R. (Eds.). (1994). *The nature of emotion: Fundamental questions*. Oxford: Oxford University Press.

Elias, N. (1978/ [1939]). *The civilizing process: Vol. 1 The history of manners*. Oxford: Basil Blackwell.

Elias, N. (1982/ [1939]). *The civilizing process: Vol. 2. State formation and civilization.* Oxford: Basil Blackwell.

Emirbayer, M., & Goldberg, C. A. (2005). Pragmatism, Bourdieu, and collective emotions in contentious politics. *Theory and Society, 34,* 469–518.

Engle, K. J. (2007). Putting mourning to work: Making sense of 9/11. *Theory, Culture & Society, 24,* 61–88.

Eppert, C. (2003). Histories re-imagined, forgotten and forgiven: Student responses to Toni Morrison's *Beloved. Changing English, 10,* 185–194.

Eppert, C. (2006). *The "wisdom of forgiveness": South African testimonial literature and reflections on witnessing as curriculum.* Paper presented at the annual meeting of the American Educational Research Association, San Francisco.

Epstein, J. (2001). Remember to forget: The problem of traumatic cultural memory. In J. Epstein & L. H. Lefkovitz (Eds.), *Shaping losses: Cultural memory and the Holocaust* (pp. 186–204). Chicago: University of Illinois Press.

Epstein, J., & Lefkovitz, L. H. (2001). Introduction: Shaping losses, cultural memory, and the Holocaust. In J. Epstein & L. H. Lefkovitz (Eds.), *Shaping losses: Cultural memory and the Holocaust* (pp. 1–12). Chicago: University of Illinois Press.

Fanon, F. (1980). *Toward the African revolution* (H. Chevalier, Trans.). London: Writers and Readers.

Farrell, K. (1998). *Post-traumatic culture: Inquiry and interpretation in the nineties.* Baltimore: Johns Hopkins University Press.

Faulkner, W. (1951). *Requiem for a nun.* New York: Random House.

Felman, S., & Laub, D. (1992). *Testimony: Crisis of witnessing in literature, psychoanalysis, and history.* New York: Routledge.

Fentress, J., & Wickham, C. (1992). *Social memory.* London: Blackwell.

Feuerverger, G. (2001). *Oasis of dreams: Teaching and learning peace in a Jewish-Palestinian village in Israel.* New York: RoutledgeFalmer.

Fierke, K. M. (2002). The liberation of Kosovo: Emotion and the ritual reenactment of war. *Focaal: European Journal of Anthropology, 39,* 93–116.

Fierke, K. M. (2004). Whereof we can speak, thereof we must not be silent: Trauma, political solipsism and war. *Review of International Studies, 30,* 471–491.

Fisher, R. J. (1997). *Interactive conflict resolution.* Syracuse, NY: Syracuse University Press.

Forrest, P. (2006). Collective guilt; individual shame. *Midwest Studies in Philosophy, 30,* 145–153.

Forter, G. (2003). Against melancholia: Contemporary mourning theory, Fitzgerald's *The Great Gatsby,* and the politics of unfinished grief. *Differences: A Journal of Feminist Cultural Sciences, 14,* 134–170.

Fortier, A.-M. (2005). Pride politics and multiculturalist citizenship. *Ethnic and Racial Studies, 28,* 559–578.

Foucault, M. (1983). The subject and power: Afterword. In H. Dreyfus & P. Rabinow (Eds.), *Michel Foucault: Beyond structuralism and hermeneutics* (pp. 208–227). Chicago: University of Chicago Press.

Foucault, M. (1990a). *The history of sexuality: Vol. 1. An introduction*. New York: Vintage Books.

Foucault, M. (1990b). *The history of sexuality: Vol 2. The use of pleasure*. New York: Vintage Books.

Foucault, M. (1990c). *The history of sexuality: Vol. 3. The care of the self.* New York: Vintage Books.

Foucault, M. (2000). *The essential works of Foucault: Power, Vol.3* (J. Faubion, Ed. ; R. Hurley et al., Trans.). New York: The New Press

Foucault, M. (2003). *Society must be defended: Lectures at the College de France, 1975–1976* (A. I. Davidson, M. Bertani, A. Fontana, & F. Ewald, Eds.; D. Macey, Trans.). New York: Picador.

Francis, D. J. (2006). Linking peace, security and developmental regionalism: Regional economic and security integration in Africa. *Journal of Peace-building and Development, 2*(3), 7–20.

Fraser, N., & Honneth, A. (Eds.). (2003). *Redistribution or recognition? A political-philosophical exchange*. New York: Verso.

Freud, S. (1917). Mourning and melancholia. In J. Strachey (Ed.), *The standard edition of the complete psychological works of Sigmund Freud* (Vol. 14, pp. 243–258). London: Hogarth.

Frow, J. (1998). The politics of stolen time. *Meaning, 57,* 351–367.

Füredi, F. (2002). *Culture of fear: Risk-taking and the morality of low expectations.* London: Continuum.

Gabbard, G. O., & Winer, J. A. (1994). Hate in the analytic setting. *Journal of the American Psychoanalytic Association, 42,* 219–231.

Gallagher, T. (2004). *Education in divided societies.* London: Palgrave/ Macmillan.

Galtung, J., & Jacobsen, C. (2000). *Searching for peace.* Sterling, VA: Pluto Press.

Galuya, G. (2006). The aesthetics of simplicity: Yang's *sadness* and the melancholic community. *Journal of Intercultural Studies, 27,* 83–100.

Game, A., & Metcalfe, A. (1996). *Passionate sociology.* London: Sage.

Gandhi, L. (2006). *Affective communities: Anticolonial thought, fin-de-siècle radicalism, and the politics of friendship*. Durham, NC: Duke University Press.

Gedi, N., & Elam, Y. (1996). Collective memory—What is it? *History and Memory, 8,* 30–50.

Gee, J. (1992). *The social mind: Language, ideology and social practice.* New York: Bergin & Garvey.

Gellner, E. (1997). *Nationalism.* New York: New York University Press.

Ghanem, A. (1998). State and minority in Israel: The case of ethnic state and the predicament of its minority. *Ethnic and Racial Studies, 21,* 428–448.

Gilroy, P. (2004). *Postcolonial melancholia.* New York: Columbia University Press.

Giroux, H. (1992). *Border crossings: Cultural workers and the politics of education.* New York: Routledge.

Giroux, H. (1997). *Pedagogy and the politics of hope: Theory, culture and schooling.* Boulder, CO: Westview Press.

Giroux, H. (2003). *The abandoned generation: Democracy beyond the culture of fear.* New York: Palgrave Macmillan.

Giroux, H.. (2004a). Critical pedagogy and the postmodern/modern divide: Towards pedagogy of democratization. *Teacher Education Quarterly, 31,* 132–153.

Giroux, H. (2004b). *The terror of neoliberalism: Authoritarianism and the eclipse of democracy.* Boulder, CO: Paradigm.

Glassner, B. (1999). *The culture of fear: Why Americans are afraid of the wrong things.* New York: Basic Books.

Goleman, D. (1995). *Emotional intelligence.* New York: Bantam Books.

Goleman, D. (1998). *Working with emotional intelligence.* New York: Bantam Books.

Gordon, A. (1997). *Ghostly matters: Haunting and the sociological imagination.* Minneapolis: University of Minnesota Press.

Gordon, P. (1999). *Face to face: Therapy as ethics.* London: Constable.

Grasso, L. (2002). *The artistry of anger: Black and white women's literature in America, 1820–1860.* Chapel Hill: University of North Carolina Press.

Gregoriou, Z. (2005). *Before the educational violence of commemoration: Derrida and the philosophical work of mourning.* Paper presented at the annual meeting of the Philosophy of Education Society of Great Britain, London.

Grossberg, L. (1988). Postmodernity and affect: All dressed up with no place to go. *Communications, 10,* 271–293.

Hadjipavlou-Trigeorgis, M. (1998). Different relationships to the land: Personal narratives, political implications and future possibilities. In V. Calotychos (Ed.), *Cyprus and its people: Nation, identity, and experience in an unimaginable community, 1955–1997* (pp. 251–276). Boulder, CO: Westview Press.

Halbwachs, M. (1950/1992). *On collective memory* (L. Coser, Trans.). Chicago: University of Chicago Press.

Hall, S. (1996). Introduction: Who needs "identity"? In S. Hall and P. du Gay (Eds.), *Questions of cultural identity* (pp. 1–17). London: Sage.

Halpern, J. (2001). *From detached concern to empathy: Humanizing medical practice.* Oxford: Oxford University Press.

Halpern, J., & Weinstein, H. M. (2004). Rehumanizing the other: Empathy and reconciliation. *Human Rights Quarterly, 26,* 561–583.

Hames-Garcia, M. R. (2000). Who are our people? Challenges for a theory of social identity. In P. Moya & M. R. Hames-Garcia (Eds.), *Reclaiming identity* (pp. 102–132). Berkeley: University of California Press.

Handelman, D. (1990). *Models and mirrors: Towards an anthropology of public events.* Cambridge, UK: Cambridge University Press.

Harding, J., & Pribram, E. D. (2004). Losing our cool? Following Williams and Grossberg on emotions. *Cultural Studies, 18,* 863–883.

Harré, R. (Ed). (1986). *The social construction of emotions.* New York: Basil Blackwell.

Harré, R., & Gillett, G. (1994). *The discursive mind.* London: Sage.
Harris, I. (1996). Editor's introduction. *Peabody Journal of Education, 71*(3), 1–11.
Harris, I. (1999). Types of peace education. In A. Raviv, L. Oppenheimer, & D. Bar-Tal (Eds.), *How children understand war and peace* (pp. 299–317). San Franscisco: Jossey-Bass.
Harris, I. (2002a). Challenges for peace educators at the beginning of the 21st century. *Social Alterantives, 21*(1), 28–31.
Harris, I. (2002b). Conceptual underpinnings of peace education. In G. Salomon & B. Nevo (Eds.), *Peace education: The concept, principles and practices around the world* (pp. 15–26). New York: Lawrence Erlbaum.
Harris, I. (2004). Peace education theory. *Journal of Peace Education, 1,* 5–20.
Harris, I. (2007). Peace education in a violent culture. *Harvard Educational Review, 77,* 350–354.
Harris, I., & Morrison, M. (2003). *Peace education* (2nd ed.). Jefferson, NC: McFarland.
Harris, I., & Synott, J. (2002). Peace education for a new century [Special issue]. *Social Alternatives, 21*(1).
Herman, J. (1997). *Trauma and recovery.* New York: Basic Books.
Hesford, W. (2004). Documenting violations: Rhetorical witnessing and the spectacle of distant suffering. *Biography, 27,* 104–144.
Hill, J. (2000). *Becoming a cosmopolitan: What it means to be a human in the new millennium.* Lanham, MD: Rowman & Littlefield.
Hobsbawm, E. J. (1983). The invention of tradition. In E. J. Hobsbawm & T. Ranger (Eds.), *The invention of tradition* (pp. 1–14). Cambridge, UK: Cambridge University Press.
Hochschild, A. R. (1975). The sociology of feeling and emotion: Selected possibilities. In M. Millman & R. M. Kanter (Eds.), *Another voice* (pp. 280–307). New York: Anchor.
Hochschild, A. R. (1983). *The managed heart: Commercialization of human feeling.* Berkeley: University of California Press.
Holmes, M. (2004a). Feeling beyond rules: Politicizing the sociology of emotion and anger in feminist politics. *European Journal of Social Theory, 7,* 209–227.
Holmes, M. (2004b). Introduction. The importance of being angry: Anger in political life. *European Journal of Social Theory, 7,* 123–132.
Horowitz, E., & Wanstrom, J. (2003). *Patriotism and political socialization: Young adults' perspectives on the anniversary of 9/11.* Paper presented at the annual meeting of the American Association for Public Opinion Research, Nashville, TN.
Hoy, D. (2004). *Critical resistance: From poststructuralism to post-critique.* Cambridge, MA: MIT Press.
Hubbard, P. (2003). Fear and loathing at the multiplex: Everyday anxiety in the post-industrial city. *Capital and Class, 80,* 51–76.

Human Rights and Equal Opportunity Commission. (1997). *Bringing them home: Report of the national inquiry into the separation of Aboriginal and Torres Strait islander children from their families.* Canberra: Commonwealth of Australia.

Humphrey, M. (2000). From terror to trauma: Commissioning truth for national reconciliation. *Social Identities, 6,* 7–27.

Ibrahim, A. (2007). Linking Marxism, globalization, and citizenship education: Toward a comparative and critical pedagogy post 9/11. *Educational Theory, 57,* 89–103.

Ignatieff, M. (1995). *Blood and belonging: Journeys into the new nationalism.* New York: Farrar, Straus & Giroux.

Ignatieff, M. (1997). *The warrior's honor: Ethnic war and the modern conscience.* New York: Henry Holt.

Jaggar, A. (1989). Love and knowledge: Emotion in feminist epistemology. In S. Bordo & A. Jaggar (Eds.), *Gender/body/knowledge: Feminist reconstructions of being and knowing* (pp. 145–171). New Brunswick, NJ: Rutgers University Press.

Kansteiner, W. (2002). Finding meaning in memory: A methodological critique of collective memory studies. *History and Theory, 41,* 179–197.

Kansteiner, W. (2004). Genealogy of a category mistake: A critical intellectual history of the cultural trauma metaphor. *Rethinking History, 8,* 193–221.

Kaplan, A. (2005). *Trauma culture: The politics of terror and loss in media and literature.* New Brunswick, NJ: Rutgers University Press.

Karstedt, S. (2002). Emotions and criminal justice. *Theoretical Criminology, 6,* 299–317.

Kemmis, S., & McTaggart, R. (Eds.). (1988). *The action research planner* (3rd ed.). Geelong, Victoria: Deakin University Press.

Kenan, S. (2005). Reconsidering peace and multicultural education after 9/11: The case of educational outreach for Muslim sensitivity curriculum in New York city. *Educational Sciences: Theory & Practice, 5,* 172–180.

Kilby, J. (2002). Redeeming memories: The politics of trauma and history. *Feminist Theory, 3,* 201–210.

Kizilyürek, N. (1993). *Cyprus beyond nation.* Nicosia, Cyprus: Cassoulides.

Kizilyürek, N. (1999). National memory and Turkish-Cypriot textbooks. *Internationale Schulbuchforschung, 21,* 387–396.

Kizilyürek, N. (2001). History textbooks and nationalism. In C. Koulouri (Ed.), *Teaching the history of Southeastern Europe* (pp. 68–73). Thessaloniki: Center for Democracy and Reconciliation in Southeast Europe.

Klein, M. (with Riviere, J.). (1962). *Love, hate and reparation.* London: Hogarth.

Kleinman, A., Das, V., & Lock, M. (Eds.). (1997). *Social suffering.* Berkeley: University of California Press.

Kleinman, A., & Kleinman, J. (1991). Suffering and its professional transformation: Toward an ethnography of interpersonal experience. *Culture, Medicine, and Psychiatry, 15,* 275–301.

Kofman, E. (2005). Citizenship, migration and the reassertion of national identity. *Citizenship Studies, 9,* 453–467.

Kosicki, P. H. (2007). Sites of aggressor-victim memory: The Rwandan genocide, theory and practice. *International Journal of Sociology, 37,* 10–29.

Krapp, P. (2005). Amnesty: Between an ethics of forgiveness and the politics of forgetting. *German Law Journal, 6,* 185–195.

Kreuzer, P. (2002). *Applying theories of ethno-cultural conflict and conflict resolution to collective violence in Indonesia.* Frankfurt, Germany: Peace Research Institute of Frankfurt.

Kriesberg, L. (1998). Coexistence and the reconciliation of communal conflicts. In E. Weiner (Ed.), *The handbook of interethnic coexistence* (pp. 182–198). New York: Continuum.

Kristeva, J. (1993). *Nations without nationalism* (L. S. Roudiez, Trans.). New York: Columbia University Press.

Kristjansson, K. (2004). Empathy, sympathy, justice and the child. *Journal of Moral Education, 33,* 291–305.

Kumashiro, K. (2002). *Troubling education: Queer activism and antioppressive pedagogy.* New York: RoutledgeFalmer.

LaCapra, D. (1998). *History and memory after Auschwitz.* Ithaca, NY: Cornell University Press.

LaCapra, D. (2001). *Writing history, writing trauma.* Baltimore: Johns Hopkins University Press.

Laclau, E. (1990). *New reflections on the revolution of our time.* London: Verso.

Leavitt, J. (1996). Meaning and feeling in the anthropology of the emotions. *American Ethnologist, 23,* 514–539.

Lederach, J. P. (1997). *Building peace: Sustainable reconciliation in divided societies.* Washington, DC: U.S. Institute of Peace.

Levi, P. (1989). *The drowned and the saved* (R. Rosenthal, Trans.). London: Abacus.

Levinas, E. (1969). *Totality and infinity: An essay on exteriority* (A. Lingis, Trans.). Pittsburgh, PA: Duquesne University Press.

Levinas, E. (1987a). *Collected philosophical papers* (A. Lingis, Trans.). Dordrecht, The Netherlands: Martinus Nijhoff.

Levinas, E. (1987b). *Time and the other and additional essays* (R. A. Cohen, Trans.). Pittsburgh, PA: Duquesne University Press.

Levinas, E., & Kearney, R. (1986). Dialogue with Emmanuel Levinas. In R. A. Cohen (Ed.), *Face to face with Levinas* (pp. 13–33). Albany: State University of New York Press.

Leys, R. (2000). *Trauma: A genealogy.* Chicago: Chicago University Press.

Lingard, B., & Mills, M. (2007). Pedagogies making a different: Issues of social justice and inclusion. *International Journal of Inclusive Education, 11,* 233–244.

Lingis, A. (2000). *Dangerous emotions.* Berkeley: University of California Press.

Lockhat, R., & Niekerk, A. V. (2000). South African children: A history of adversity violence and trauma. *Ethnicity and Health, 5,* 291–302.

Loizos, P. (1981). *The heart grown bitter: The chronicle of Cypriot war refugees.* Cambridge, UK: Cambridge University Press.

Loizos, P. (1988). Intercommunal killing in Cyprus. *Man, 23,* 639–653.

Loizos, P. (1998). How might Turkish and Greek Cypriots see each other more clearly? In V. Calotychos (Ed.), *Cyprus and its people: Nation, identity, and experience in an unimaginable community, 1955–1997* (pp. 35–52). Boulder, CO: Westview Press.

Loizos, P. (2005). *Ethnic pluralism and co-habitation in Cyprus: A socio-historical review.* Paper presented at the conference on the "Psycho-Social Aspects of Reunification," Intercollege, Nicosia, Cyprus.

Lorde, A. (1984). *Sister outsider: Essays and speeches.* Trumansburg, NY: Crossing Press.

Lupton, D. (1998). *The emotional self: A sociocultural exploration.* London: Sage.

Lutz, C. (1988). *Unnatural emotions: Everyday sentiments on a Micronesian atoll and their challenge to Western theory.* Chicago: University of Chicago Press.

Lutz, C., & Abu-Lughod, L. (Eds.). (1990). *Language and the politics of emotion.* Cambridge, UK: Cambridge University Press.

Lyman, P. (1981). The politics of anger: On silence, ressentiment and political speech. *Socialist Review, 11*(3), 55–74.

Lyman, P. (2004). The domestication of anger: The use and abuse of anger in politics. *European Journal of Social Theory, 7,* 133–147.

Lyotard, J.-F. (1998). *The differend: Phrases in dispute* (G. V. D. Abbeele, Trans.). Minneapolis: University of Minnesota Press.

Lyotard, J.-F., & Thébaud, J.-L. (1985). *Just gaming* (W. Godzich, Trans.). Minneapolis: University of Minnesota Press.

Mack, J. (1990). The psychodynamics of victimization among national groups in conflict. In V. Volkan, D. Julius, & J. Montville (Eds.), *The psychodynamics of international relationships: Vol. 1. Concepts and theories* (pp. 119–129). Toronto, Canada: Lexington Books.

Mann, M. (2004). *The dark-side of democracy.* Cambridge, UK: Cambridge University Press.

Martusewicz, R. (2001). *Seeking passage: Post-structuralism, pedagogy, ethics.* New York: Teachers College Press.

Massumi, B. (2002). *Parables for the virtual: Movement, affect, sensation.* Durham, NC: Duke University Press.

Matsuda, M. J. (1993). Public response to racist speech: Considering the victim's story. In M. J. Matsuda, C. R. Lawrence, & R. Delgado (Eds.), *Words that wound: Critical race theory, assaultive speech and the first amendment* (pp. 17–52). Boulder, CO: Westview Press.

Mavratsas, C. (1996). Approaches to nationalism: Basic theoretical considerations in the study of the Greek-Cypriot Case and a historical overview. *Journal of the Hellenic Diaspora, 22,* 77–102.

Mavratsas, C. (1999). National identity and consciousness in everyday life: Toward a sociology of knowledge of Greek-Cypriot nationalism. *Nations and Nationalism, 5,* 91–104.
McCormack, D. P. (2003). An event of geographical ethics in spaces of affect. *Transactions of the Institute of British Geographers, 28,* 488–507.
McGlynn, C., Niens, U., Cairns, E., & Hewstone, M. (2004). Moving out of conflict: The contribution of integrated schools in Northern Ireland to identity, attitudes, forgiveness and reconciliation. *Journal of Peace Education, 1,* 147–163.
McKnight, A. (2004). Historical trauma, the persistent of memory and the pedagogical problems of forgiveness, justice and peace. *Educational Studies, 36,* 140–158.
McLaren, P. (1994). Critical pedagogy, political agency, and the pragmatics of justice: The case of Lyotard. *Educational Theory, 44,* 319–340.
McLaren, P. (1989/2003). *Life in schools: An introduction to critical pedagogy in the foundations of education* (4th ed.). Boston: Allyn and Bacon.
Mead, G. H. (1934). *Mind, self and society.* Chicago: University of Chicago Press.
Meltzer, B., & Musolf, G. R. (2002). Resentment and ressentiment. *Sociological Inquiry, 72,* 240–255.
Mendus, S. (1989). *Toleration and the limits of liberalism.* New York: McMillan.
Merton, R. (1976). *Sociological ambivalence and other essays.* New York: Free Press.
Mestrovic, S. G. (1997). *Postemotional society.* London: Sage.
Metz, J. B. (1972). The future in the memory of suffering. In J. B. Metz (Ed.), *New questions on God* (pp. 9–25). New York: Herder and Herder.
Metz, J. B. (1980). *Faith in history and society: Towards a fundamental practical theology.* New York: Seabury.
Metz, J. B. (1999). In the pluralism of religious and cultural worlds: Notes toward a theological and political program. *Cross Currents, 49,* 227–236.
Miall, H., Ramsbotham, O., & Woodhouse, T. (2000). *Contemporary conflict resolution: The prevention, management, and transformation of deadly conflicts.* Cambridge, UK: Polity Press.
Middleton, D., & Edwards, D. (Eds.). (1990). *Collective remembering.* London: Sage.
Mohanty, C. (1997). *Literary theory and the claims of history: Postmodernism, objectivity, multicultural politics.* Ithaca, NY: Cornell University Press.
Mohanty, S. (2000). The epistemic status of cultural identity: On *Beloved* and the postcolonial condition. In P. M. L. Moya & M. R. Hames-Garcia (Eds.), *Reclaiming identity: Realist theory and the predicament of postmodernism* (pp. 29–66). Berkeley: University of California Press.
Montgomery, K. E. (2005). Banal race-thinking: Ties of blood Canadian history textbooks and ethnic nationalism. *Paedagogica Historica, 41,* 313–336.
Moore, D., & Aweiss, S. (2002). Hatred of “others” among Jewish, Arab, and Palestinian students in Israel. *Analyses of Social Issues and Public Policy, 2,* 151–172.

Morris, J. A., & Feldman, D. C. (1997). Managing emotions in the workplace. *Journal of Managerial Issues, 9,* 257–274.

Moya, P., & Hames-Garcia, M. R. (Eds.). (2000). *Reclaiming identity: Realist theory and the predicament of postmodernism.* Berkeley: University of California Press.

Naas, M. (2003). History's remains: Of memory, mourning, and the event. *Research in Phenomenology, 33,* 75–96.

Nancy, J.-L. (1991). *The inoperative community.* Minneapolis: University of Minnesota Press.

Nash, M. (2005). How to be thankful for being free: Searching for a convergence of discourses on teaching patriotism, citizenship, and United States history. *Teachers College Record, 107,* 214–240.

Nguyen, Minh T. (2000). It matters to get the facts straight. In P. Moya & M. R. Hames-Garcia (Eds.), *Reclaiming identity: Realist theory and the predicament of postmodernism* (pp. 171–204). Berkeley: University of California Press.

Nieto, S. (2000). Placing equity front and center: Some thoughts on transforming teacher education for a new century. *Journal of Teacher Education, 51,* 180–187.

Nora, P. (1989). Between memory and history: Les lieux de memoire. *Representations, 26,* 7–25.

Olick, J. K. (1999). Collective memory: The two cultures. *Sociological Theory, 17,* 333–348.

Olick, J. K., & Robbins, J. (1998). Social memory studies: From "collective memory" to the historical sociology of mnemonic practices. *Annual Review of Sociology, 24,* 105–140.

Oliver, K. (2000). Witnessing otherness in history. In H. Marchitello (Ed.), *What happens to history: The renewal of ethics in contemporary thought* (pp. 41–66). New York: Routledge.

Oliver, K. (2001). *Witnessing: Beyond recognition.* Minneapolis: University of Minnesota Press.

Ophir, A. (2000). The identity of the victims and the victims of identity: A critique of Zionist ideology for a pot-Zionist age. In L. J. Silberstein (Ed.), *Mapping Jewish identities* (pp. 174–201). New York: New York University Press.

Opotow, S. (1990). Moral exclusion and injustice: An overview. *Journal of Social Issues, 46*(1), 1–20.

Opotow, S. (1995). Drawing the line: Social categorization, moral exclusion, and the scope of justice. In B. B. Bunker & J. Z. Rubin (Eds.), *Conflict, cooperation, and justice* (pp. 347–369). San Francisco: Jossey-Bass.

Opotow, S. (2001). Reconciliation in times of impunity: Challenges for social justice. *Social Justice Research, 14,* 149–170.

Opotow, S., Gerson, J., & Woodside, S. (2005). From moral exclusion to moral inclusion: Theory for teaching peace. *Theory into Practice, 44,* 303–318.

Ortloff, D. (2006). Becoming European: A framing analysis of three countries' civics education curricula. *European Education, 37,* 35–49.

Ostovich, S. T. (2002). Epilogue: Dangerous memories. In A. Confino & P. Fritzsche (Eds.), *The work of memory: New directions in the study of German society and culture* (pp. 239–256). Urbana, IL: University of Illinois Press.

Ostovich, S. T. (2005). Dangerous memories and reason in history. *KronoScope, 5,* 41–57.

Paez, D., Basabe, N., & Gonzalez, J. L. (1997). Social processes and collective memory: A cross-cultural approach to remembering political events. In J. W. Pennebaker, D. Paez, & B. Rimé (Eds.), *Collective memory of political events: A sociological perspective* (pp. 147–174). Mahwah, NJ: Lawrence Erlbaum Associates.

Papadakis, Y. (1995). Nationalist imaginings of war in Cyprus. In R. Hinde & H. Watson (Eds.), *War, a cruel necessity? The bases of institutionalized violence* (pp. 54–67). London: I. B. Tauris.

Papadakis, Y. (1998). Greek Cypriot narratives of history and collective identity: Nationalism as a contested process. *American Ethnologist, 25,* 149–165.

Papastergiadis, N. (2006). Ambient fears. In M. Margaroni and E. Yiannopoulou (Eds.), *Metaphoricity and the politics of mobility* (pp. 119–140). New York: Rodopi.

Parkin, D. (1986). Toward an apprehension of fear. In D. L. Scruton (Ed.), *Sociophopbics: The anthropology of fear* (pp. 158–172). Boulder, CO: Westview Press.

Pensky, M. (1989). On the use and abuse of memory. *Philosophy and Social Criticism, 15,* 352–378.

Pertersen, R. D. (2002). *Understanding ethnic violence: Fear, hatred, and resentment in twentieth century eastern Europe.* Cambridge, UK: Cambridge University Press.

Petonito, G. (2000). Racial discourse and enemy construction. In P. G. Coy & L. M. Woehrle (Eds.), *Social conflicts and collective identities* (pp. 20–40). Oxford: Rowman & Littlefield.

Popkewitz, T. S., & Brennan, M. (Eds.). (1998). *Foucault's challenge: Discourse, knowledge, and power in education.* New York: Teachers College Press.

POST Research Institute. (2004). *Education for peace: Pilot application for the history and literature books of the 5th grade of the elementary school.* Nicosia, Cyprus: POST.

POST Research Institute. (2007). *Textual and visual analyses of the lower secondary school history textbooks: Comparative analysis of the old and new history textbooks.* Nicosia, Cyprus: POST.

Potter, J., & Wetherell, M. (1990). *Discourse and social psychology.* London: Sage.

Probyn, E. (2004). Shame in the habitus. *Sociological Review, 52,* 224–248.

Probyn, E. (2005). *Blush: Faces of shame.* Minneapolis, MN: University of Minnesota Press.

Rachman, S. (1998). *Anxiety.* Hove, UK: Psychology Press.

Rae, T. (2003). *Dealing with feeling: An emotional literacy curriculum.* Newport, UK: Positively MAD.

Rak, J. (2003). Do witness: *Don't: A Woman's Word* and trauma as pedagogy. *Topia, 10,* 53–71.

Ramanathapillai, R. (2006). The politicizing of trauma: A case study of Sri Lanka. *Peace and Conflict: Journal of Peace Psychology, 12,* 1–18.

Redding, P. (1999). *The logic of affect.* Melbourne: Melbourne University Press.

Reddy, W. (2001). *The navigation of feeling: A framework for the history of emotions.* Cambridge, UK: Cambridge University Press.

Reicher, S., & Hopkins, N. (2001). *Self and nation.* London: Sage.

Remarque, E. M. (1982). *All quiet on the western front.* New York: Ballantine Books.

Retzinger, S. (1991). *Violent emotions.* Newbury Park, CA: Sage.

Ricoeur, P. (1999). Memory and forgetting. In R. K. Earney & M. D. Ooley (Eds.), *Questioning ethics: Contemporary debates in philosophy* (pp. 5–11). New York: Routledge.

Ricoeur, P. (2004). *Memory, history, forgetting* (K. Blamey and D. Pellauer, Trans.). Chicago: University of Chicago Press.

Robin, C. (2004). *Fear: The history of a political idea.* New York: Oxford University Press.

Ropers-Huilman, B. (1999). Witnessing: Critical inquiry in a poststructural world. *Qualitative Studies in Education, 12,* 21–35.

Rorty, R. (1989). *Contingency, irony, and solidarity.* Cambridge, UK: Cambridge University Press.

Rose, N. (1990). *Governing the soul: The shaping of the private self.* London: Routledge.

Rose, N. (1998). *Inventing ourselves: Psychology, power and personhood.* Cambridge, UK: Cambridge University Press.

Rose, N. (1999). *Powers of freedom: Reframing political thought.* Cambridge, UK: Cambridge University Press.

Rothman, J. (1998). Dialogue in conflict: Past and future. In E. Weiner (Ed.), *The handbook of interethnic coexistence* (pp. 217–235). New York: Continuum.

Rothstein, R. L. (Ed.). (1999). *After the peace: Resistance and reconciliation.* Boulder, CO: Lynne Rienner.

Runnymede Trust. (2000). *The future of multi-ethnic Britain: The Parekh report.* London: Profile Books.

Salaita, S. (2005). Ethnic identity and imperative patriotism: Arab Americans before and after 9/11. *College Literature, 32,* 146–168.

Salomon, G. (2004). A narrative-based view of coexistence education. *Journal of Social Issues, 60,* 237–287.

Salomon, G., & Nevo, B. (2001). The dilemmas of peace education in intractable conflicts. *Palestine—Israel Journal, 8,* 64–78.

Salomon, G., & Nevo, B. (Eds.). (2002). *Peace education: The concept, principles, and practices around the world.* Mahwah, NJ: Lawrence Erlbaum Associates.

Sampson, E. (1993). *Celebrating the other: A dialogic account of human nature.* Hertfordshire, UK: Harvester Wheatsheaf.

Sant Cassia, P. S. (2006). Guarding each other's dead, mourning one's own: The problem of missing persons and missing pasts in Cyprus. *South European Society and Politics, 11,* 111–128.

Santos, M. S. (2001). Memory and narrative in social theory: The contributions of Jacques Derrida and Walter Benjamin. *Time & Society, 10,* 163–189.

Savage, J. (2004). Researching emotion: The need for coherence between focus, theory and methodology. *Nursing Inquiry, 11,* 25–34.

Scheff, J. T. (1992). Emotions and identity: A theory of ethnic nationalism. In C. Calhoun (Ed.), *Social theory and the politics of identity* (pp. 277–303). Oxford: Blackwell.

Scheff, J. T. (1994). *Bloody revenge: Emotions, nationalism and war.* Boulder, CO: Westview Press.

Scheff, J. T., & Retzinger, S. (1991). *Emotions and violence: Shame and rage in destructive conflicts.* Lexington, MA: Lexington.

Schwartz, S., & Sagiv, L. (1995). Identifying culture-specifics in the content and structure of values. *Journal of Cross-Cultural Psychology, 26*(1), 92–116.

Schweber, S. A. (2004). *Making sense of the Holocaust: Lessons from classroom practice.* New York: Teachers College Press.

Scott, S. (2001). Surviving selves: Feminism and contemporary discourse of child sexual abuse. *Feminist Theory, 2,* 349–361.

Sedwick-Kosofsky, E. (2003). *Touching feeling: Affect, pedagogy, performativity.* Durham, NC: Duke University Press.

Sedwick-Kosofsky, E., & Sedwick, A. F. (Eds.). (1995). *Shame and its sisters: A Silvan Tomkins reader.* Durham, NC: Duke University Press.

Selzer, M. (1997). Wound culture: Trauma in the pathological public sphere. *October, 80,* 3–26.

Sharp, P. (2001). *Nurturing emotional literacy: A practical guide for teachers, parents and those in the caring professions.* London: David Fulton.

Shechter, H., & Salomon, G. (2005). Does vicarious experience of suffering affect empathy for an adversary? The effects of Israelis' visits to Auschwitz on their empathy for Palestinians. *Journal of Peace Education, 2,* 125–138.

Shor, I. (1992). *Empowering education: Critical teaching for social change.* Chicago: University of Chicago Press.

Siegel, M. (2002). History is the opposite of forgetting: The limits of memory and the lessons of history in interwar France. *Journal of Modern History, 74,* 770–800.

Simmel, G. (1971). The metropolis and mental life. In D. Levine (Ed.), *Georg Simmel: On individuality and social forms* (pp. 324–339). Chicago: University of Chicago Press.

Simon, R. (1993). Forms of insurgency in the production of popular memories: The Columbus quincentenary and the pedagogy of counter-commemoration. *Cultural Studies, 7,* 73–88.

Simon, R. (2005). *The touch of the past: Remembrance, learning, and ethics.* New York: Palgrave Macmillan.

Simon, R., & Eppert, C. (1997). Remembering obligation: Pedagogy and the witnessing of testimony of historical trauma. *Canadian Journal of Education, 22,* 175–191.

Simon, R., Eppert, C., Clamen, M., & Beres, L. (2001). Witness as study: The difficult inheritance of testimony. *Review of Education/Pedagogy/Cultural Studies, 22,* 285–322.

Simon, R., Rosenberg, S., & Eppert, C. (Eds.). (2000). *Beyond hope and despair: Pedagogy and the representation of historical trauma.* Totowa, NJ: Rowman and Littlefield.

Smooha, S. (1996). Ethno-democracy: Israel as an archetype. In P. Ginosar & A. Bareli (Eds.), *Zionism: A contemporary polemic* (pp. 277–311). Jerusalem: BenGurion University (Hebrew).

Solomon, R. (1990). *A passion for justice: Emotions and the origins of the social contract.* Reading, MA: Addison-Wesley.

Sontag, S. (2003). *Regarding the pain of others.* London: Penguin Books.

Sorabji, C. (2006). Managing memories in post-war Sarajevo: Individuals, bad memories, and new wars. *Journal of the Royal Anthropological Institute, 12,* 1–18.

Spelman, E. V. (1989). Anger and insubordination. In A. Garry & M. Pearsall (Eds.), *Women, knowledge and reality: Explorations in feminist philosophy* (pp. 263–273). London: Unwin Hyman.

Spelman, E. V. (1997). *Fruits of sorrow: Framing our attention to suffering.* Boston: Beacon Press.

Spivak, G. (1987). *In other worlds: Essays in cultural politics.* New York: Routledge.

Sprinzak, D., Segev, Y., Bar, E., & Levi-Mazloum, D. (2001). *Facts and figures about education in Israel.* Jerusalem: State of Israel, Ministry of Education, Economics and Budgeting Administration.

Spyrou, S. (2001a). Being one and more than one: Greek Cypriot children and ethnic identity in the flow of everyday life. *Disclosure, 10,* 73–94.

Spyrou, S. (2001b). Those on the other side: Ethnic identity and imagination in Greek-Cypriot children's lives. In H. B. Schwartzman (Ed.), *Children and anthropology: Perspectives for the 21st century* (pp. 167–185). London: Bergin and Garvey.

Spyrou, S. (2002). Images of "the other": The Turk in Greek Cypriot children's imaginations. *Race, Ethnicity and Education, 5,* 255–272.

Spyrou, S. (2006). Constructing "the Turk" as an enemy: The complexity of stereotypes in children's everyday worlds. *South European Society and Politics, 11,* 95–110.

Starling, R. (2002). Addressing the dead: Of friendship, community and the work of mourning. *Angelaki: Journal of the Theoretical Humanities, 7,* 107–124.

Staub, E. (1999). Preventing genocide: Activating bystanders, helping victims heal, helping groups overcome hostility. In L. Chorbajian & G.

Shirinian (Eds.), *Studies in comparative genocide* (pp. 251–260). New York: St. Martin's Press.

Stearns, C. Z., & Stearns, P. N. (1986). *Anger: The struggle for emotional control in America's history.* Chicago: University of Chicago Press.

Stephen, M. (2000). Witchcraft, grief, and the ambivalence of emotions. *American Ethnologist, 26,* 711–737.

Stocker, M. (1996). *Valuing emotions.* Cambridge, UK: Cambridge University Press.

Stovel, L. (2000). *Confronting ethnic chauvinism in a post-war environment: NGOs and peace education in Bosnia.* Bradford, UK: Centre for Conflict Resolution Working Paper series.

Streich, G. (2002). Is there a right to forget? Historical injustices, race, memory, and identity. *New Political Science, 24,* 525–542.

Suga, K. (2006). *Echos-Monde* and abrasions: Translation as a form of dialogue. *Social Identities, 12,* 17–28.

Sutherland, M. (2002). Educating citizens in Europe. *European Education, 34,* 77–95.

Svašek, M. (2002). The politics of emotions: Emotional discourses and displays in post-cold war contexts. *Focaal- European Journal of Anthropology, 39,* 9–27.

Svašek, M. (2005). The politics of chosen trauma: Expellee memories, emotions and identities. In K. Milton & M. Svašek (Eds.), *Mixed emotions: Anthropological studies of feeling* (pp. 195–214). Oxford: Berg.

Swaine, L. (1996). Blameless, constructive, and political anger. *Journal for the Theory of Social Behavior, 26,* 257–274.

Tajfel, H., & Turner, J. C. (1986). The social identity theory of intergroup behavior. In S. Worchel and L. W. Austin (Eds.), *Psychology of intergroup relations* (pp. 2–24). Chicago: Nelson-Hall

Tajfel, H., & Turner, J. C. (2001). An integrative theory of intergroup conflict. In M. A. Hogg & D. Abrams (Eds.), *Intergroup relations: Essential readings* (pp. 94–114). Ann Arbor, MI: Edwards Brothers.

Tarnopolsky, C. (2004). Prudes, perverts, and tyrants: Plato and the contemporary politics of shame. *Political Theory, 32,* 468–494.

Tawil, S. (Ed.). (1997). *Final report and case studies of the workshop on educational destruction and reconstruction in disrupted societies.* Geneva, Switzerland: UNESCO-International Bureau of Education.

Theodossopoulos, D. (2006). Introduction: The "Turks" in the imagination of the "Greeks." *South European Society & Politics, 11,* 1–32.

Thrift, N. (2004). Intensities of feeling: Towards a spatial politics of affect. *Geografiska Annaler, 86*(2), 55–76.

Thrift, N. (2005). But malice aforethought: Cities and the natural history of hatred. *Transactions of the Institute of British Geographers, 30,* 133–150.

Todd, S. (2007). Promoting a just education: Dilemmas of rights, freedom and justice. *Educational Philosophy and Theory, 39,* 592–603.

Todorov, T. (2003). *Hope and memory: Reflections on the twentieth century.* London: Atlantic Books.

Tracy, S. J. (2000). Becoming a character for commerce: Emotion labor, self-subordination, and discursive construction of identity in a total institution. *Management Communication Quarterly, 14,* 90–128.

Van Peer, C. (2006). Education on populations matters in Europe: Results from a comparative survey among students in five European countries. *Compare, 36,* 105–123.

Varenne, H., & McDermott, R. (1998). *Successful failure: The schools America builds.* Boulder, CO: Westview Press.

Venn, C. (2005). The repetition of violence: Dialogue, the exchange of memory, and the question of convivial socialities. *Social Identities, 11,* 283–298.

Verducci, S. (2000). A conceptual history of empathy and a question it raises for moral education. *Educational Theory, 50,* 63–80.

Vertzberger, Y. I. (1997). The antinomies of collective political trauma: A pre-theory. *Political Psychology, 18,* 863–876.

Volkan, V. (1979). *Cyprus-war and adaptation: A psychoanalytic history of two ethnic groups in conflict.* Charlottesville: University Press of Virginia.

Volkan, V. (1988). *The need to have enemies and allies: From clinical practice to international relationships.* Northvale, NJ: Jason Aronson.

Volkan, V. (1997). *Bloodlines: From ethnic pride to ethnic terrorism.* Boulder, CO: Westview Press.

Volkan, V. (2001). Transgenerational transmissions and chosen traumas: An aspect of large-group identity. *Group Analysis, 34,* 79–97.

Volkan, V., & Itzkowitz, N. (1994). *Turks and Greeks: Neighbors in conflict.* Cambridgeshire, UK: Eothen Press.

Waller, J. (1992). *Becoming evil: How ordinary people commit genocide and mass killing.* Oxford: Oxford University Press.

Wang, H. (2005). Aporias, responsibility, and the im/possibility of teaching multicultural education. *Educational Theory, 55,* 45–59.

Watt, I. (1997). *Myths of modern individualism: Faust, Don Quixote, Don Juan, Robinson Crusoe.* Cambridge, UK: Cambridge University Press.

Weigert, A. (1991). *Mixed emotions: Certain steps toward understanding ambivalence.* Albany: State University of New York Press.

Weigert, A., & Franks, D. D. (1989). Ambivalence: A touchstone of the modern temper. In D. Franks & E. Doyle McCarthy (Eds.), *The sociology of emotions: Original essays and research papers* (pp. 205–227). Greenwich, CT: JAI Press.

Weingarten, K. (2003). *Common shock—witnessing violence every day: How we are harmed, how we can heal.* New York: Dutton.

Weis, L., & Fine, M. (2001). Extraordinary conversations in public schools. *Qualitative Studies in Education, 14,* 497–523.

Welch, S. (1985). *Communities of resistance and solidarity: A feminist theology of liberation.* Maryknoll, NY: Orbis Books.

Wetherell, M., & Potter, J. (1992). *Mapping the language of racism.* New York: Harverster Wheatsheaf.

White, G. M. (2000). Emotional remembering: The pragmatics of national memory. *Ethos, 27,* 505–529.

Williams, B. (1993). *Shame and necessity.* Berkeley: University of California Press.

Williams, R. (1961). *Culture and society, 1780–1950.* Hardmondsworth, UK: Penguin Books.

Williams, S. (2001). *Emotion and social theory: Corporeal reflections on the (ir)rational.* London: Sage.

Winter, J. (2006). *Remembering war: The Great War between memory and history in the twentieth century.* New Haven, NJ: Yale University Press.

Winter, J., & Sivan, E. (1999). Setting the framework. In J. Winter & E. Sivan (Eds.), *War and remembrance in the twentieth century* (pp. 6–39). Cambridge, UK: Cambridge University Press.

Wintersteiner, W., Spajić-Vrkaš, V., & Teutsch, R. (Eds.). (2003). *Peace education in Europe: Vision and experiences.* New York: Waxman.

Wohlwend, K. (2006). "I am not an American girl": Resisting discourses of agency, innocence and patriotism. *International Journal of Equity and Innovation in Early Childhood, 4,* 48–62.

Woodward, K. (2000). Traumatic shame: Toni Morrison, televisual culture, and the cultural politics of the emotions. *Cultural Critique, 46,* 210–240.

Woodward, K. (2005). Calculating compassion. In L. Berlant (Ed.), *Compassion: The culture and politics of an emotion* (pp. 59–86). New York: Routledge.

Worsham, L. (2001). Going postal: Pedagogic violence and the schooling of emotion. In H. Giroux and K. Myrsiades (Eds.), *Beyond the corporate university: Culture and pedagogy in the new millennium* (pp. 229–265). New York: Rowman & Littlefield.

Yanay, N. (1995). The meaning of hatred as narrative: Two versions of an experience. *Journal of Narrative and Life History, 5,* 353–368.

Yanay, N. (1996). National hatred, female subjectivity, and the boundaries of cultural discourse. *Symbolic Interaction, 19,* 21–36.

Yanay, N. (2002a). Hatred as ambivalence. *Theory, Culture & Society, 19*(3), 71–88.

Yanay, N. (2002b). Understanding collective hatred. *Analyses of Social Issues and Public Policy, 2,* 53–60.

Young, J. (1993). *The texture of memory: Holocaust memorials and meaning.* New Haven, CT: Yale University Press.

Zembylas, M. (2005). *Teaching with emotion: A postmodern enactment.* Greenwich, CT: Information Age Publishing.

Zembylas, M. (2006a). *Suffering, memory and forgiveness: Derrida, Levinas and the pedagogical challenges of reconciliation.* Paper presented at the annual meeting of the American Educational Research Association, San Francisco.

Zembylas, M. (2006b). Witnessing in the classroom: The ethics and politics of affect. *Educational Theory, 56,* 305–324.

Zembylas, M. (2007). *Five pedagogies, a thousand possibilities: Affect, critical hope, transformation.* Rotterdam, The Netherlands: SensePublishers.

Zembylas, M., & Boler, M. (2002). On the spirit of patriotism: Challenges of a "pedagogy of discomfort." *Teachers College Record On-line*. Retrieved December 16, 2007, from http://tcrecord.org.

Zembylas, M., & Fendler, L. (2007). Reframing emotion in education through lenses of *parrhesia* and *care of the self. Studies in Philosophy and Education, 26,* 319–333.

Zembylas, M., & Karahasan, H. (2006). The politics of memory and forgetting in pedagogical practices: Towards pedagogies of reconciliation and peace in divided Cyprus. *Cyprus Review, 18*(2), 15–35.

Zerubavel, E. (1996). Social memories: Steps to a sociology of the past. *Qualitative Sociology, 19,* 283–299.

Ziarek, E. (2006). Encounters possible and impossible: Derrida and Butler on mourning. *Philosophy Today, 50*(SPEP Supplement), 144–155.

Zinner, E. S., & Williams, M. B. (1999). *When a community weeps: Case studies in group survivorship*. London: Brunner/Mazel.

Zuzovsky, R., & Yakir, R. (1999). Teaching for social change: A Palestinian-Israeli case of peace education. *Mediterranean Journal of Educational Studies, 4*(1), 67–81.

Index